Ano-Rectal Endosonography and Manometry in Paediatrics

Mario Lima • Giovanni Ruggeri
Editors

Ano-Rectal Endosonography and Manometry in Paediatrics

Springer

Editors
Mario Lima
Pediatric Surgery Unit
IRCCS Sant'Orsola University-Hospital
University of Bologna
Bologna, Italy

Giovanni Ruggeri
Pediatric Surgery Unit
IRCCS Sant'Orsola University-Hospital
University of Bologna
Bologna, Italy

ISBN 978-3-030-97667-5 ISBN 978-3-030-97668-2 (eBook)
https://doi.org/10.1007/978-3-030-97668-2

This Springer imprint is published by the registered company Springer Nature Switzerland AG
The registered company address is: Gewerbestrasse 11, 6330 Cham, Switzerland

Preface

The application of endoanal ultrasonography in children, as the use of high-resolution anorectal manometry, is not recent at all. Nevertheless, differently from what happened in adults medicine, these two sophisticated investigations are prerogative of only few centers in Europe.

Furthermore, there is an increased interest in both pediatricians and pediatric surgeons on these diagnostic technologies. This is the reasons why we felt the need to organize and lead, from Bologna, two internal workshops, respectively, in June and October 2020, with a surprising successful attendance. These two events involved some of the most important experts on congenital colorectal anomalies in Europe. The idea to write this monograph was the outcome of this successful experience; our wish was to create a starting point for future studies, research, and technical developments, with the aim of offering both young and experienced pediatric surgeons a comprehensive textbook and an updated review on this rapidly changing field. We hope that result will achieve such high goals.

The text is divided into 16 chapters. The first chapters (Chaps. 1–3) were written in order to recall some basics of anorectal anatomy and physiology, necessary to understand and interpret these sophisticated endoanal investigations. The central chapters (Chaps. 4–13 and Chap. 16) explore the present main indications of endoanal ultrasonography and manometry. The last chapters (Chaps. 14, 15) instead are meant to show some innovative applications of these technologies, opening the reader's mind on the future possibilities on this field.

I wish to thank all those who have actively collaborated on the creation of this book, and all the authors, who for friendship and desire to pass on their knowledge, agreed to provide their contribution.

Bologna, Italy Mario Lima

Contents

1 History of Endocavitary Ultrasound

Tommaso Gargano, Claudio Antonellini, Neil Di Salvo, Chiara Cordola, and Mario Lima

In the last decades ultrasound has probably become the tool with a major clinical impact allowing physicians to have several information without exposing patients to the adverse effects of radiation and providing wildly available, fast, and accurate images. For these reasons someone refers to ultrasound as the stethoscope of the twenty-first century [1]. However its history dates back to the twentieth century.

It was the Italian physician Lazzaro Spallanzani (1729–1799) in his "*Opuscoli di Fisica*" who first described the mechanism of nocturn navigation of bats using sound for orientating instead of light. That was the first description of spatial waves.

The first machine with the ability of producing sound waves was the "Galton whistle" created by Francis Galton (1822–1911) in the 1880s. In the same years Jacques and Pierre Curie (1855–1941 and 1859–1906), studying quartz crystal, discovered the piezoelectric effect known as the ability of certain crystalline materials to generate an electric charge in response to applied mechanical stress. Lately they also discovered the inverse piezoelectric effect known as the ability of the liquid crystal to produce electricity under the vibration produced by ultrasound waves.

The French physicist Paul Langevin (1872–1946) in 1916 studied the use of high-frequency sound waves for finding German submarines during the First World War, which was the first attempt of technological application of ultrasound. His studies were the base on which the first SONAR (SOund Navigation And Ranging) was later developed by the American Navy.

The credit of introducing ultrasound in medicine goes to Karl Dussik (1908–1968) who first presented the idea of using ultrasound as a diagnostic tool in the paper "Uber die moglichkeit hochfrequente mechanische schwingungen als diagnostisches hilfsmittel zu verwerten" (On the possibility of using ultrasound waves as a diagnostic aid). Later, with his brother Friedrich (1910–1988) he produced the "hyperphonography" apparatus which used ultrasound with the attempts of detecting brain tumors.

In 1956 Ian Donald (1910–1987) introduced the use of ultrasound as a diagnostic tool in gynecological and obstetrics field studying ovarian cysts and fetal growth. He used a one-dimensional machine (amplitude mode). Only in 1963, when brightness-mode devices were developed, which established the operator to visualize the images in two dimensions, the commercial use of ultrasound began.

T. Gargano (✉) · C. Antonellini · N. Di Salvo
C. Cordola · M. Lima
Pediatric Surgery Unit, IRCCS Sant'Orsola University-Hospital, University of Bologna, Bologna, Italy
e-mail: tommaso.gargano2@unibo.it; claudio.antonellini@aosp.bo.it; niel.disalvo.mebi@pa.omceo.it; mario.lima@unibo.it

M. Lima, G. Ruggeri (eds.), *Ano-Rectal Endosonography and Manometry in Paediatrics*,
https://doi.org/10.1007/978-3-030-97668-2_1

Pulsed Doppler device was designed in 1966 by Don Backer (1943–2010), Dennis Watkins, and John Reid opening new incredible possibilities in the study of blood flow.

Since then the diagnostic accuracy of the ultrasound machines and the quality of the images acquired have continuously improved bringing it to the success and wild diffusion that it has today.

Rectal endosonography was first described in 1956 [5] by Wild J. and colleagues [2] who visualized recurrent rectal cancer with an endoscopic ultrasonic probe. However, the technology had still many limitations and the quality of the images was low and these reasons prevented it from spreading in the following decades.

In 1983 Dragsted and Gammelgaard [3] presented the first series of rectal cancer staged preoperatively by intrarectal ultrasonography [4] leading the way to its use in adult patients.

In the 1980s the radiological team of St. Mark's Hospital in London (Fig. 1.1) started using in the anorectum a 360° transducer designed for studying urogenital tract pathology (BK Medical, Denmark). In 1989, their studies brought to publication the first scientific article about anal endosonography and normal ultrasonographic appearance of the anus and the rectal canal [5].

Initially, the company BK Medical was the spearhead in technology used for the assessment of anorectal pathology; afterwards, Aloka and Hitachi began producing 360° transducers for this area. The high cost of the equipment has been a major constraint in many centers, mainly in developing countries. Attempting to resolve this situation, startup companies emerged with 360° images equipment with relatively low costs, such as Echoson, Accuview, and Halo (Fig. 1.2) [4].

Indeed in anorectal pathology of adults endoanal, transperineal, and endovaginal sonography are nowadays wildly used and allow a full evaluation of the patient.

Moreover the negative side of being an operator-dependent method is now partially offset by the introduction of 3D technology in the most modern equipment.

Since 1989, endoanal ultrasonography has become an essential investigation for those physicians who deal daily with coloproctological issues, at least in adult medicine.

The main fields of use of endoanal sonography in adult patients are assessment of anal

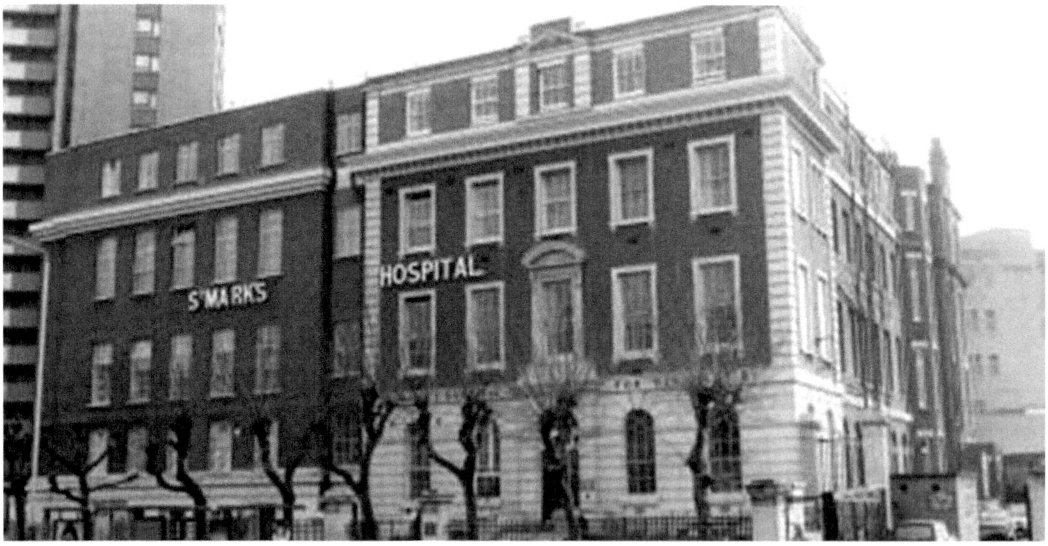

Fig. 1.1 St. Mark's Hospital. Here, in 1989, radiologists started using the same endocavitary probe urologists used for the prostate in order to define the normal ultrasonographic appearance of the anus and the rectal canal (photo taken from St. Mark's Hospital site). Courtesy of St Mark's Hospital Foundation & Academic Institute

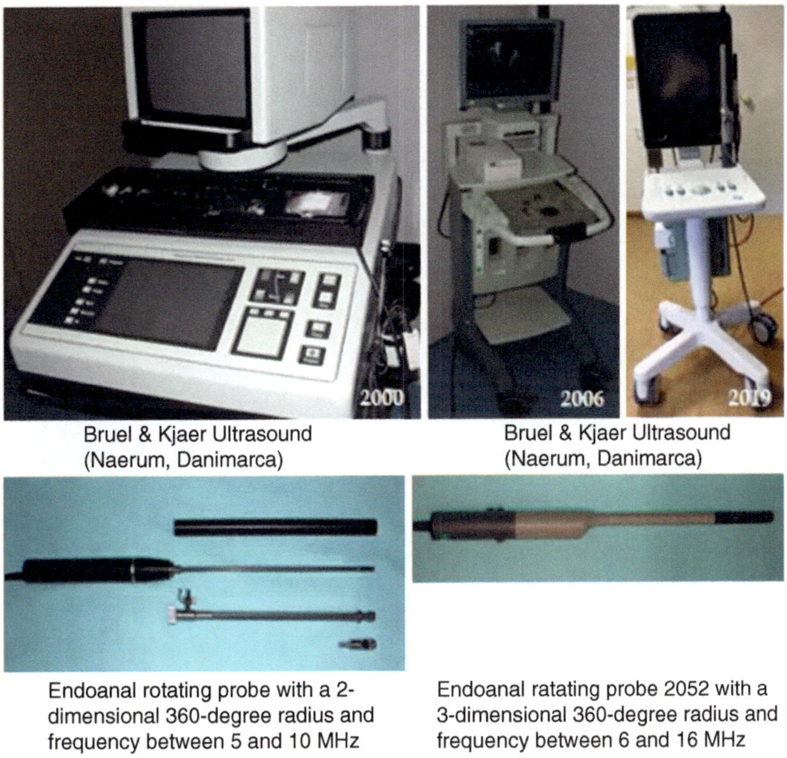

Fig. 1.2 Evolution of ultrasonographic equipment (BK Medical)

abscess and fistulas especially in those who suffer from chronic inflammatory bowel diseases, study of the pelvic and perineal structures in patients with fecal incontinence, and staging and follow-up of rectal tumors.

Transperineal and transvaginal ultrasounds are instead used for the study of pelvic floor dysfunctions.

The first use of endoanal sonography in children dates back to 1994 where the Dutch study group of Benninga studied a small population of healthy children with the aim of defining the typical characteristics of the anorectal structure in a pediatric population [6].

In the same year the Norwegian group of Emblem studied with anal endosonography adolescent patients operated for anorectal malformation and compared them to healthy children [7].

Since then the knowledge about normal echostructure of anorectum and the echographic presentation of anorectal congenital disease have been constantly increasing with multiplication of centers which began using these methods for clinical assessment in children.

References

1. Villanueva-Herrero JA, Hansen M, Alarcon-Bernes L, Jiménez-Bobadilla B. Endoanal ultrasound for anorectal diseases. Gastrointestinal diseases. SM Group; 2017. p. 1–23.
2. Wild J, Reid J. Diagnostic use of ultrasound. Br J Physiol Med. 1956;19:248.
3. Dragsted J, Gammelgaard J. Endoluminal ultrasonic scanning in the evaluation of rectal cancer: a preliminary report of 13 cases. Gastrointest Radiol. 1983;8:367–9.
4. Senagore AJ. Intrarectal and intra-anal ultrasonography in the evaluation of colorectal pathology. Surg Clin North Am. 1994;74(6):1465–73. https://doi.org/10.1016/s0039-6109(16)46492-3.
5. Law PJ, Bartram CL. Anal endosonography: technique and normal anatomy. Gastrointest Radiol. 1989;14:349–53.
6. Benninga MA, Wijers OB, van der Hoeven CW, Taminiau JA, Klopper PJ, Tytgat GN, Akkermans LM. Manometry, profilometry, and endosonography: normal physiology and anatomy of the anal canal in healthy children. J Pediatr Gastroenterol Nutr. 1994;18(1):68–77.
7. Emblem R, Diseth T, Mørkrid L, Stien R, Bjordal R. Anal endosonography and physiology in adolescents with corrected low anorectal anomalies. J Pediatr Surg. 1994;29(3):447–51. https://doi.org/10.1016/0022-3468(94)90589-4.

2 Basics of Endoanal Ultrasounds

Giovanni Ruggeri, Stefania Pavia, Neil Di Salvo, Eduje Thomas, and Mario Lima

2.1 Fundamental Principles of Ultrasound Imaging

The mechanism which leads to the production of images in the exams used nowadays is based on the interaction between the organ under examination and the energy produced by external sources, such as X-rays, ultrasound, and injected radioactive substances. Two possible scenarios arise from this physical phenomenon: the energy wave manages to pass through the organ, or it is totally or partially reflected or scattered. Both possibilities end up in the creation of different kind of images which can be visualized thanks to dedicated devices.

Although discovered before X-rays in 1880, it took longer time for ultrasounds to be applied in medicine. The first practical application of ultrasound was recorded during the World War I with the purpose of aiding the detection of submarines. Before then, echolocation had been ascertained in bats by Lazzaro Spallanzani in 1794, when he demonstrated that bats hunted and navigated by inaudible sound, not by vision. The application of the ultrasound wave in medicine began in 1950s. First introduced in obstetrics, it later gained popularity in all medical domains also thanks to the introduction of the processing of the signals of gray scale in 1974. From a merely clinical point of view, the role played by its noninvasiveness in the acceptance of the ultrasound has been paramount.

Transmission technology is based on distinguishing the tissue with different absorbance of ultrasound and rendering an image that consists of a mosaic of lighter and darker places. Instead, diagnostic ultrasound, also called sonography or diagnostic medical sonography, exploits the technique of formation of images by reflection [1].

In physics, the term "ultrasound" describes a form of energy emitted at a frequency above the limit of human audibility. Generally, audible sound has a frequency between 20 Hz and 20 kHz (cycles per second = Hertz, Hz) (Fig. 2.1). Typical diagnostic sonographic scanners operate with frequencies ranging from 2 to 20 MHz, corresponding to wavelengths of 1–0.1 mm in tissue, as the frequency of oscillations is inversely proportional to wavelength.

Ultrasound is generated by a piezoelectric crystal, commonly made by lead zirconate titanate (PZT), encased in a probe and responsible for the transmission and reception of impulses. Their frequency depends on the thickness of the crystal. The piezoelectric effect is symmetrical, so that the same crystal can be used as a receiver

G. Ruggeri · S. Pavia · N. Di Salvo (✉)
E. Thomas · M. Lima
Pediatric Surgery Unit, IRCCS Sant'Orsola University-Hospital, University of Bologna, Bologna, Italy
e-mail: stefaniavalentinamaria.pavia@aosp.bo.it; niel.disalvo.mebi@pa.omceo.it; eduje.thomas@studio.unibo.it; mario.lima@unibo.it

M. Lima, G. Ruggeri (eds.), *Ano-Rectal Endosonography and Manometry in Paediatrics*,
https://doi.org/10.1007/978-3-030-97668-2_2

Fig. 2.1 Wave spectrum

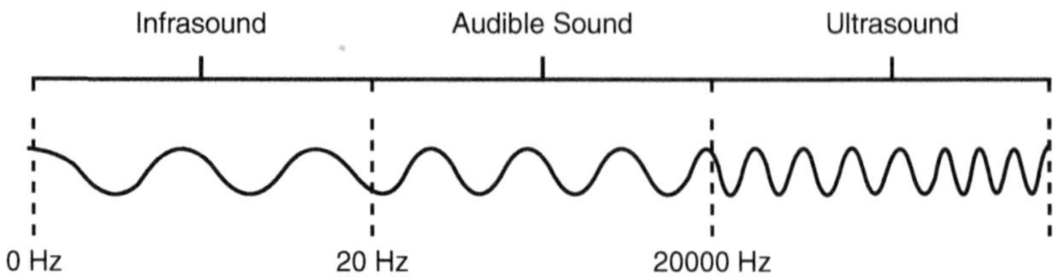

to produce small electrical signal when struck by an ultrasound wave [2].

There are different modalities of propagation in tissue and diagnostic ultrasound uses longitudinal waves. This means that particles move in the same direction as the wave, which produces a succession of rarefactions and compressions transmitted due to elastic forces between adjacent particles. Each repetition of this back-and-forth motion is called a cycle. The most important parameters describing a wave are:

- Wavelength
- Frequency
- Velocity
- Intensity

The length wave (λ) is the distance between two bands of compression or rarefaction. The transmission velocity of the energy through the medium, also called velocity of propagation (ν), depends on the strength of the elastic forces between particle (elasticity of the tissue) and the masses of the particles (density of the medium). These two factors determine the impedance of the tissue. The wavelength and the frequency are linked to the velocity of propagation by the equation $\nu = f \times \lambda$.

A sound beam running through the body is attenuated, meaning reduced in intensity, by a combination of different phenomena: absorption, reflection, refraction, and diffusion. Intensity at each point of the wave is defined as the energy flow per unit of time through the unit area perpendicular to the direction of propagation at the point considered.

The acoustic absorption is mainly due to the conversion of ultrasonic energy into thermal energy which depends on three factors: (1) frequency, (2) viscosity of the conducting medium, and (3) relaxation time of the medium defined as the time necessary for a molecule to return to its original position after its initial displacement. The absorption coefficient varies from tissue to tissue and is expressed as a fraction between the amplitude (measured in decibels) and the centimeter of tissue depth for each megahertz.

The energy conveyed by the ultrasonic wave is also yielded to the receiving transducer when reflected or refracted away from the returning line of sight or if the beam diverges. Reflection technology (echo) registers the pulse reflected from the boundary between two tissues with different acoustic impedance. The entity of the returning echo depends on the difference in the acoustic impedance and on the angle of incidence of the beam. Within soft tissues, only a small percentage of the beam is reflected, whereas almost total reflection occurs at tissue/gas interfaces. The behavior of the ultrasound wave in relation to the angle of incidence obeys the Snell's law for light: waves are reflected at an angle equal to the angle of incidence (specular reflection) (Fig. 2.2). According to such rule, the optimal reflection is obtained when the wave and the reflecting surface are perpendicular. This implies that using a rotating endoprobe within circular structures yields the best return as most specular reflections will be at right angles.

A different phenomenon, known as scattering, can be observed when the irregularities in the surface are of the same order of size as the wavelength: each small interface is vibrated by the mechanical insult it has received from the ultrasound pulse and the vibration is re-emitted equally in all directions. Scattering within biological tissues reduces the propagating energy so that it contributes to attenuation together with reflection and absorption [3].

According to the principle of energy wave attenuation, the echoes from deep structures are weaker

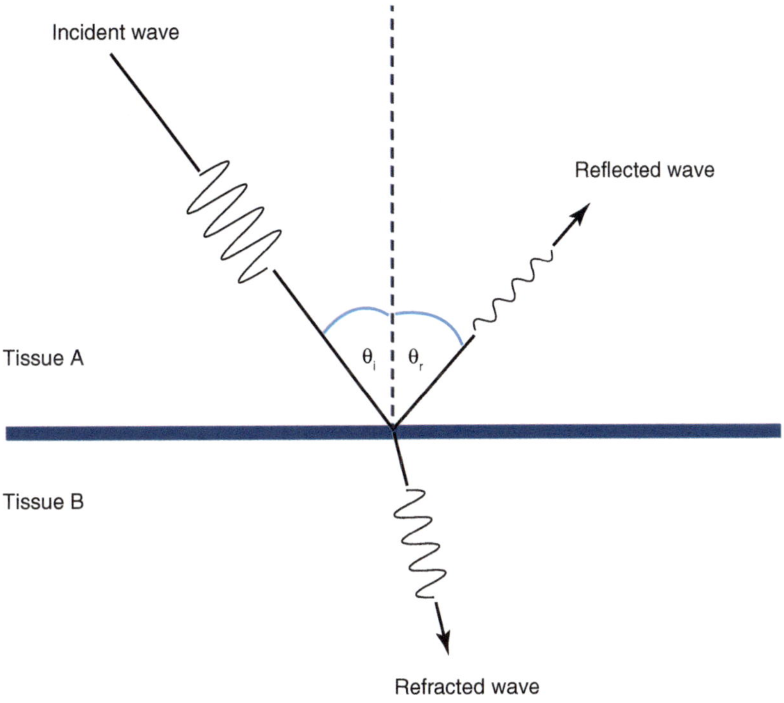

Fig. 2.2 Reflection and refraction of ultrasound waves

and poorly detected compared to those from closer tissues. Therefore, an accurate adjustment of amplification is important to enable adequate screen display, which is achieved by applying progressively increasing amplification (gain) to later echoes in proportion to their depth using a time-varying amplifier. However, these adjustments alter both axial (capacity to distinguish two-point reflectors in the direction of the axis of the ultrasound beam) and lateral resolution (capacity to detect two-point reflectors at an equal distance from the transducer but situated in two different directions from it).

The transducer used to produce ultrasound beams is also responsible for their detection thanks to the piezoelectric crystal, which converts electrical signals into mechanical vibrations and vice versa. The signal is then transmitted to a series of electronic circuits and commonly rendered as a luminous spot of the screen of an oscilloscope. The image is extrapolated by correlating the luminous point on the screen and the distance traveled by the ultrasonic wave [4].

Four different modes of ultrasound are used in medical imaging [1]:

1. A-mode: It is the simplest form of ultrasound, in which a single transducer scans a line through the body with the echoes plotted on screen as a function of depth.
2. B-mode: In B-mode ultrasound, a linear array of transducers simultaneously scans a plane through the body that can be viewed as a 2D image on screen.
3. M-mode: M stands for motion. In this mode, a rapid sequence of B-mode scans run in succession enables doctors to see and measure the range of motion, as the organ boundaries that produce reflections move relative to the probe.
4. Doppler mode: This mode makes use of Doppler effect in measuring and visualizing blood flow, in terms of not only direction, but also its relative velocity.

2.2 Endoanal Ultrasonography

Transrectal, transvaginal, and transperineal ultrasonography have been widely accepted as popular imaging modalities to study anorectal pathology in adults. In the pediatric setting these diagnostic tools remain exclusive to few third-level centers worldwide.

The first attempts to apply endoanal ultrasonography to children, dating back to 1994. On the one hand the Dutch group of Benninga studied a small cohort of normal children aiming to define the exact endosonographic anorectal anatomy in the pediatric age. On the other hand, Emblem et al. in Norway examined the muscular anorectal components of adolescents with anorectal malformation (ARM) corrected in childhood [5, 6].

Since then, giant strides have been made in improving the understanding and definition of the features of the normal echostructure of the anal canal and the alterations which occur in congenital anorectal malformations.

Endoanal ultrasonography was first proposed in the 1960s for the evaluation of rectal tumors. However, the device used provided low-quality images. Towards the end of the 1980s, a group of radiologists at St. Mark's in London implemented the use of a 360° transducer designed for urogenital diseases in the study of anorectal pathologies [7].

At the current state of art, the transducers used to perform endoanal ultrasonography operate in a range between 5 and 15 MHz with a focal distance of 2.8–6.2 cm. Higher frequencies provide better resolution with clear definition of rectal wall layers, whereas lower frequencies are essential for the assessment of enlarged lymph nodes and perirectal tissues [8]. The transducer is made by a crystal which rotates at 4–6 cycles per second to construct a 360° image, encased in endoprobes with a minimum diameter of 7 mm. The cylindric nature of the anal canal favors the 360° axial view at right angles to the lumen obtained with mechanically rotated endoprobes (Figs. 2.3 and 2.4).

Two types of transducers have been commercialized. The first contains a circular array of crystals which moves along the axis of the probe, granting the possibility of acquiring two-dimensional 360° representation before a proper three-dimensional reconstruction. In the three-dimensional acquisition system, the probe automatically moves back and forth over 6 cm. The second, designed more recently, entails a linear array of crystals which rotates covering an angle up to 360°, to provide three-dimensional pictures and a better orientation during the visualization of the area of interest. This second probe has the benefit of rendering 2D images on a linear plane.

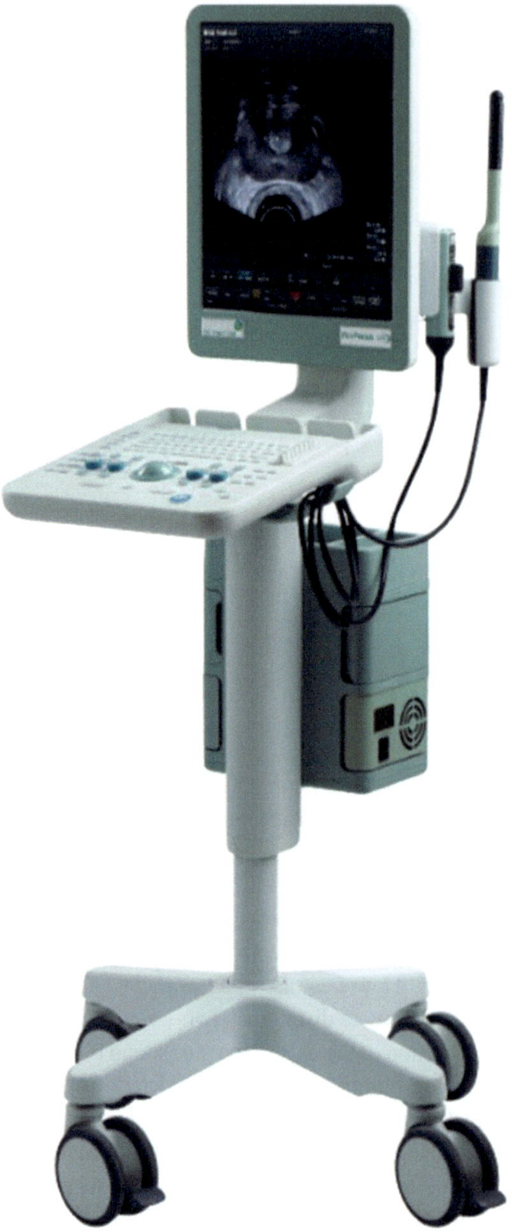

Fig. 2.3 Endoanal Ultrasonography device

The rendering software elaborates a series of axial images obtained automatically each at 0.05 mm to create a 3D volume displayed as a cube (Fig. 2.5). Such representation conceals several benefits, as the 3D image can be freely rotated, rendered, tilted, and sliced to allow the operator to better visualize the lesion at different angles to get the most information out of the data.

Fig. 2.4 Endoanal Ultrasonography probe (diameter: 17 mm, circular array, high frequency up to 16 MHz)

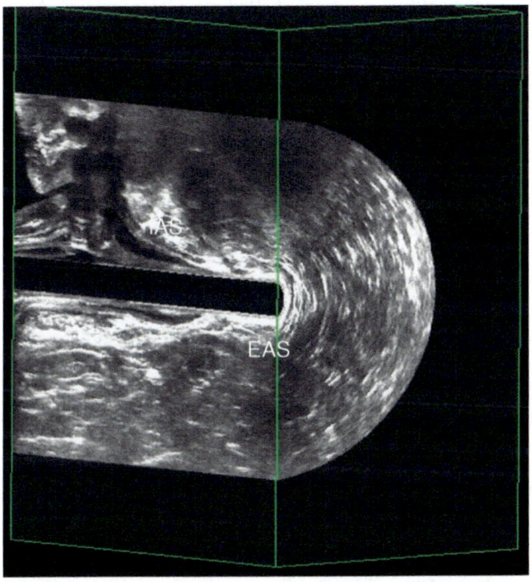

Fig. 2.5 Three-dimensional reconstruction of the sphincteric complex: puborectalis muscle, internal anal sphincter (IAS), and external anal sphincter (EAS)

In case of perianal fistulas, it is possible to perform contrast-enhanced endoanal ultrasonography using peroxide as the contrast agent.

At last, the three-dimensional integration reduces operator-dependent errors to the minimum.

Endoanal ultrasonography of the anal canal can be performed with the patient in the left lateral decubitus position with their knees bent at 90°, in prone or supine lithotomy position.

The cone is filled with degassed water for acoustic coupling. Endoanal probes require the use of a latex condom that serves as a water bath in the evaluation of the anal canal and sphincter complex. The procedure can be conducted without bowel preparation, but the use of cleansing enemas may optimize imaging. Sedation is not required before the procedure in the adult patient but may be necessary in a pediatric setting. Probe is introduced into the rectum, aligned in standard orientation, and then slowly retracted down the anal canal until the hyperechoic puborectal muscle used as a landmark is seen.

In contrast-enhanced endoanal ultrasonography, after examination of the perianal skin to identify external opening, a conventional EAUS is performed, the transducer is then removed, and external openings are cannulated. The transducer is then reinserted and EAUS is performed during gentle injection of peroxide through the cannula.

2.3 Normal Anatomy of the Anal Canal

The anal canal is 2–4 cm long, extending from the anorectal junction to the dentate line. The circular smooth muscle of the rectal wall continues caudally as the internal anal sphincter (IAS). The voluntary muscle from the levator ani and the puborectalis muscle, conjoined in a muscular cylinder, form the external anal sphincter (EAS) which encircles the IAS. The intersphincteric space, between EAS and IAS, is occupied by fat and muscle fibers originating from the outer longitudinal component of the rectal wall and the striated muscle fibers from the levator ani. The levator ani, subdivided into the iliococcygeus, pubococcygeus, and puborectalis (the most medial, forming a "U-shaped" sling behind the anorectum) muscles, represents the fundamental structure of the pelvic floor.

As it is made up by smooth muscle fibers, the IAS is supplied by autonomic nerves; it contrib-

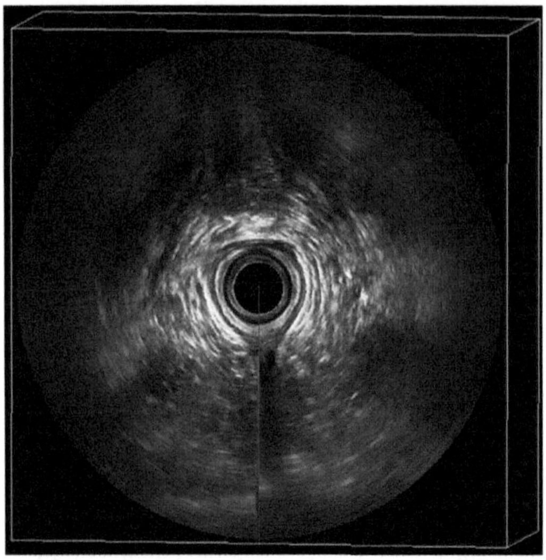

Fig. 2.6 Middle third of the anal canal: the IAS appears as a hypoechoic ring, whereas the EAS appears as a heterogeneously hyperechoic ring

utes about 85% of the resting pressure and is chiefly responsible for anal continence at rest. Instead, the EAS is innervated by the pudendal nerve, a mixed nerve which subserves both sensory and motor function [9] (Fig. 2.6).

These are the ultrasonographic features of the described structures [10]:

- Puborectalis muscle: it appears as a U-shaped hyperechoic band.
- IAS: it appears homogeneously hypoechoic; it does not extend beyond subcutaneous external sphincter.
- The intersphincteric space: it appears moderately echogenic, due to the presence of fibrous stroma.
- EAS: it usually appears hyperechoic and has heterogeneous appearance; the variability in echogenicity relates to variations in the orientation of some of its fibers. Unlike males, in 98% of women the EAS is shorter anteriorly compared to posteriorly.

References

1. Carovac A, Smajlovic F, Junuzovic D. Application of ultrasound in medicine. Acta Inform Med. 2011;19(3):168–71. https://doi.org/10.5455/aim.2011.19.168-171.
2. Santoro GA, Di Falco G. Atlas of endoanal and endorectal ultrasonography. Springer; 2004.
3. Rumack CM, Levine D. Diagnostic ultrasound, 2-volume set. Elsevier—Health Science Division; 2017.
4. Santoro GA, Di Falco G. Benign anorectal diseases: diagnosis with endoanal and endorectal ultrasound and new treatment options. Springer; 2006.
5. Benninga MA, Wijers OB, van der Hoeven CW, Taminiau JA, Klopper PJ, Tytgat GN, Akkermans LM. Manometry, profilometry, and endosonography: normal physiology and anatomy of the anal canal in healthy children. J Pediatr Gastroenterol Nutr. 1994;18(1):68–77.
6. Emblem R, Diseth T, Mørkrid L, Stien R, Bjordal R. Anal endosonography and physiology in adolescents with corrected low anorectal anomalies. J Pediatr Surg. 1994;29(3):447–51. https://doi.org/10.1016/0022-3468(94)90589-4.
7. Law PJ, Bartram CI. Anal endosonography: technique and normal anatomy. Gastrointest Radiol. 1989;14(4):349–53. https://doi.org/10.1007/BF01889235.
8. Kim MJ. Transrectal ultrasonography of anorectal diseases: advantages and disadvantages. Ultrasonography. 2015;34(1):19–31. https://doi.org/10.14366/usg.14051.
9. De la Portilla F, López-Alonso M. Endosonography of the anal canal: findings in children. Dis Colon Rectum. 2009;52(4):711–4. https://doi.org/10.1007/DCR.0b013e31819ed12c.
10. Thakar R, Sultan AH. Anal endosonography and its role in assessing the incontinent patient. Best Pract Res Clin Obstet Gynaecol. 2004;18(1):157–73. https://doi.org/10.1016/j.bpobgyn.2003.09.007.

3 Functional Anatomy of the Pelvic Floor and the Anorectum

Vincenzo D. Catania, Beatrice Randi, Neil Di Salvo, and Mario Lima

3.1 Pelvic Floor

The pelvic floor (pelvic diaphragm) is a funnel-shaped musculotendinous termination of the pelvic outlet. This multifunctional unit has connections to the bony pelvis, to organs, and to the extensive fibroelastic network in the fat containing anatomical spaces.

Pelvic floor muscles have two major functions; they provide (1) support or act as a "floor" for the abdominal viscera including the rectum and (2) constrictor or continence mechanism to the urethral, anal, and vaginal orifices (in females) [1–4].

The bony pelvis is composed of sacrum, ileum, ischium, and pubis. It is divided into the false (greater) and true (lesser) pelvis by the pelvic brim: the sacral promontory, the anterior ala of the sacrum, the arcuate line of the ilium, the pectineal line of the pubis, and the pubic crest that culminates in the symphysis pubis [1, 5].

Although often thought of as a single muscular layer only (the pelvic diaphragm or levator plate), the pelvic floor is in fact more complex[6].

The pelvic floor constitutes four principal layers, which are, from superior to inferior, the **endopelvic fascia**, the **muscular pelvic diaphragm** (commonly referred to as the levator ani muscle, which is constituted by three different muscles: *puborectalis, pubococcygeus, and iliococcygeus*), the **perineal membrane** (urogenital diaphragm), and a superficial layer of muscles with a primary role of support being the **superficial transverse perinei** [5, 7, 8].

The pelvic floor is traversed by the urethra and anal sphincters, and in women by the vagina (Fig. 3.1) [7].

The urogenital diaphragm, also called the triangular ligament, is a strong, muscular membrane that occupies the area between the symphysis pubis and ischial tuberosities and stretches across the triangular anterior portion of the pelvic outlet. The urogenital diaphragm is external and inferior to the pelvic diaphragm.

The pelvic ligaments are not classic ligaments but are thickenings of retroperitoneal fascia and consist primarily of blood and lymphatic vessels, nerves, and fatty connective tissue [2, 9].

The nerve supply to the pelvic floor and related organs is by branches of the sacral plexus—the pudendal nerve (coursing inferior to the pelvic floor), the levator ani nerve (coursing superior to the pelvic floor), and the parasympathetic pelvic splanchnic nerves (nervi erigentes) and the sympathetic supply by the hypogastric nerve. Higher regulating levels of the central nervous system (e.g., pontine micturition center, cerebral cortex) are crucial for proper function of the pelvic floor [10].

V. D. Catania (✉) · B. Randi · N. Di Salvo · M. Lima
Pediatric Surgery Unit, IRCCS Sant'Orsola University-Hospital, University of Bologna, Bologna, Italy
e-mail: vincenzodavide.catania@aosp.bo.it; beatrice.randi@aosp.bo.it; niel.disalvo.mebi@pa.omceo.it; mario.lima@unibo.it

M. Lima, G. Ruggeri (eds.), *Ano-Rectal Endosonography and Manometry in Paediatrics*, https://doi.org/10.1007/978-3-030-97668-2_3

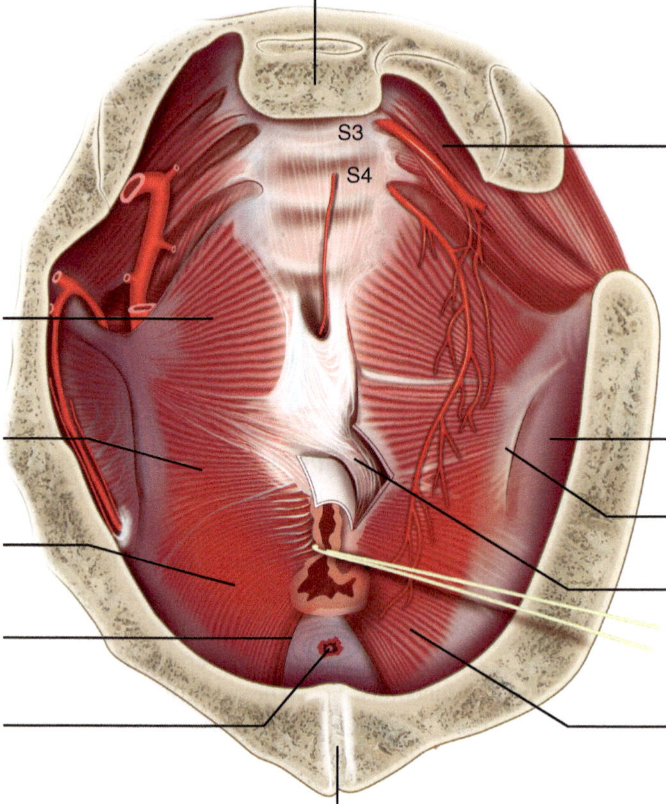

Fig. 3.1 Pelvic floor muscles, pelvic view This work is in the public domain because it was published in the United States between 1926 and 1963, and although there may or may not have been a copyright notice, the copyright was not renewed. For further explanation, see Commons:Hirtle chart and the copyright renewal logs. Note that it may still be copyrighted in jurisdictions that do not apply the rule of the shorter term for US works (depending on the date of the author's death), such as Canada (50 years p.m.a.), Mainland China (50 years p.m.a., not Hong Kong or Macao), Germany (70 years p.m.a.), Mexico (100 years p.m.a.), Switzerland (70 years p.m.a.), and other countries with individual treaties 1918, Philadelphia and New York; Publisher: Lea and Febiger
Henry Gray: Gray's Anatomy, The Anatomy of Human Body (20th edition), https://upload.wikimedia.org/wikipedia/commons/b/be/Grant_1962_209.png

3.2 Endopelvic Fascia

The endopelvic fascia is an adventitial layer covered by parietal peritoneum on top of the pelvic diaphragm and visceral structures. This fascia forms a continuous sheetlike mesentery, extending from the uterine artery at its cephalic margin to the point at which the vagina fuses with the levator ani muscles below [1, 11]. It is attached to the diaphragmatic part of the pelvic fascia along the tendinous arch, and has been subdivided in accordance with the viscera to which it is related. Thus its anterior part, known as the **vesical layer,** forms the anterior and lateral ligaments of the bladder. Its middle part crosses the floor of the pelvis between the rectum and vesiculae seminales as the **rectovesical layer** in males. Otherwise, in females, the median part of the endopelvic fascia is perforated by the vagina and it constitutes the attachments between the uterus and the rectum (*paracolpium* and *parametrium*). Its posterior portion passes to the side of the rectum; it forms a loose sheath for the rectum, but is firmly attached around the anal canal; this portion is known as the **rectal layer** [2, 8].

This fascia is important for passive support of visceral organs and pelvic floor and has expansive properties.

3.3 Pelvic Diaphragm

In 1555, Andreas Vesalius was the first who described the pelvic floor muscles as a unique complex, which he named "musculus sedem attollens" [1].

The definition of pelvic diaphragm has been introduced by Meyer (1861). This muscular complex included primitive flexors and abductors of the caudal part of the vertebral column. These muscles included *ischiococcygeus*, *iliococcygeus*, and *pubococcygeus* and these three muscles were felt to constitute the **levator ani muscle**. Holl (1897) and Thompson in 1899 described that some of the pubococcygeus muscle fibers, instead of inserting into the coccyx, looped around the rectum and to these fibers they assigned the name "puborectalis" or "sphincter recti" (Figs. 3.2 and 3.3) [1, 7, 12].

The puborectalis muscle is now included in the levator ani muscle group and the term "levator ani" is used synonymously with pelvic dia-

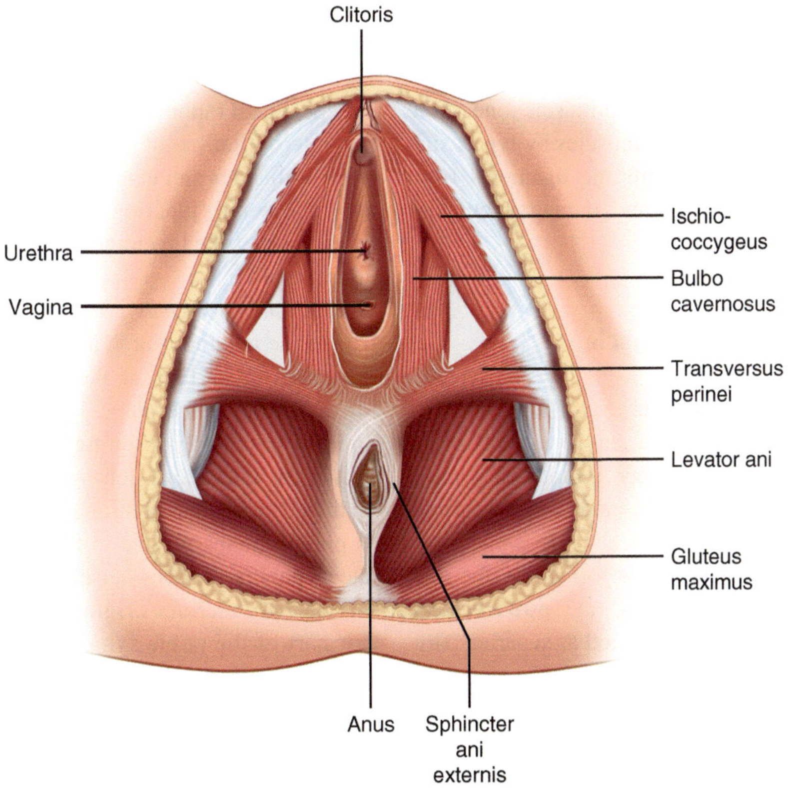

Fig. 3.2 Levator ani muscle, perineal view This image is in the public domain because it is a mere mechanical scan or photocopy of a public domain original, or—from the available evidence—is so similar to such a scan or photocopy that no copyright protection can be expected to arise. The original itself is in the public domain for the following reason: 1918, Philadelphia and New York; Publisher: Lea and Febiger

Henry Gray: Gray's Anatomy, The Anatomy of Human Body (20th edition), https://upload.wikimedia.org/wikipedia/commons/7/7d/Levator_ani.png

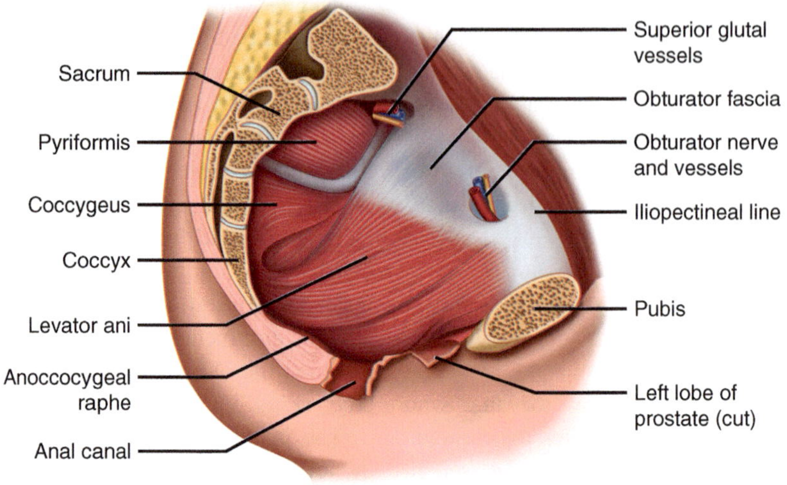

Fig. 3.3 Levator ani muscle, lateral view This image is in the public domain because it is a mere mechanical scan or photocopy of a public domain original, or—from the available evidence—is so similar to such a scan or photocopy that no copyright protection can be expected to arise. The original itself is in the public domain for the following reason: 1918, Philadelphia and New York; Publisher: Lea and Febiger **Henry Gray**: Gray's Anatomy, The Anatomy of Human Body (20th edition), https://upload.wikimedia.org/wikipedia/commons/6/63/Gray404.png

phragm muscles. The constant muscle tone of the levator ani and coccygeus muscles by type I striated muscle fibers prevents the ligaments from becoming overstretched and damaged by constant tension [13].

3.3.1 Levator Ani Muscle

The major function of the levator ani muscle is supporting and raising the pelvic visceral structures. It also helps in proper *sexual functioning, defecation,* and *urination*, and allows various structures to pass through it.

The levator ani muscle is formed by the **iliococcygeus**, **pubococcygeus**, and **puborectalis**. These muscles can be identified as separate parts by their origin and direction. The *iliococcygeus* muscle and *pubococcygeus* muscle arise from the ischial spine, the tendinous arch of the levator ani muscle, and the pubic bone. They originate from the pectinate line of the pubic bone and the fascia of the obturator internus muscle and are inserted into the coccyx [7].

The *iliococcygeus* arises from the posterior half of the tendinous arch inserting into the last two segments of the coccyx and the midline anococcygeal raphe. The anococcygeal raphe is the interdigitation of the iliococcygeal fibers from both sides and extends from the coccyx to the anorectal junction. The iliococcygeus forms a sheetlike layer and is often largely aponeurotic [14].

The *pubococcygeus* arises from the anterior half of the tendinous arch and the periosteum of the posterior surface of the pubic bone at the lower border of the pubic symphysis, its fibers directed posteriorly inserting into the anococcygeal raphe and coccyx. It has variation depending on the gender, where it inserts to either the vagina as pubovaginalis muscle in females or the prostate as puboprostaticus muscle in males [9].

The *puborectalis* originates from the inferior part of the pubic symphysis and the superior fascial layer of the urogenital diaphragm. The puborectalis forms a U-shaped sling around the urogenital hiatus and meets the fibers from the opposite muscle at the midline. Contraction of the puborectalis lifts and compresses the urogenital hiatus. The puborectalis is the main part of this muscle and goes around the upper part of the anus where it is attached posteriorly to the anococcygeal ligament. It plays an important role in maintaining an acute recto-anal angle crucial for a proper gross fecal continence [1, 7, 12].

The blood supply to the levator ani muscle comes from different branches of the inferior gluteal artery, inferior vesical artery, and pudendal artery. Venous drainage and lymphatic drainage are along the corresponding veins accompanying the arteries.

The primary nerve supply is from the nerves originating from the S3 and S4, with fibers coming from the S2 and coccygeus plexus. The pubococcygeus and the puborectalis muscles are primarily innervated by pudendal nerve branches such as the inferior rectal nerve while the iliococcygeus is primarily innervated by the sacral S3 and S4 nerves [1, 2, 7, 9, 15].

3.3.2 Coccygeus Muscle

The coccygeus forms the posterior part of the pelvic diaphragm. This shelflike triangular musculotendinous structure has its origin at the ischial spine and courses along the posterior margin of the internal obturator muscle. The muscles insert at the lateral side of the coccyx and the lowest part of the sacrum. The sacrospinous ligament is at the posterior edge of the coccygeus muscle and is fused with this muscle. The proportions of the muscular and ligamentous parts may vary. The coccygeus is not part of the levator ani, having a different function and origin.

The coccygeus muscle is innervated by the third and fourth sacral spinal nerves on its superior surface [7, 16].

3.3.3 Perineal Membrane

In the anterior pelvis, below the levator ani muscles, there is a dense triangular shaped fibromuscular membrane called the **perineal membrane** [17]. The perineal membrane lies at the level of the hymen and attaches the urethra, vagina, and perineal body to the ischiopubic rami. It is a set of uni-layer musculofascial structure that spans the arch between the ischiopubic rami with each side attached to the other through their connection in the perineal body [14]. The compressor urethrae and urethrovaginal sphincter muscles are associated with the upper surface of the perineal membrane constituting its muscular component. The structure bridges the gap between the inferior pubic rami bilaterally and the perineal body.

3.3.4 Superficial Transverse Perinei

The **superficial transverse perinei**, the *bulbospongiosus*, and the *ischiocavernosus* are the external genital muscles and form the most superficial components of the pelvic floor. The superficial transverse perinei has a supportive function, and the bulbospongiosus and the ischiocavernosus play a role in urinary continence and sexual function [2, 5]. The superficial transverse perinei spans the posterior edge of the perineal membrane and inserts at the external sphincter and perineal body in women while in men at the perineal point.

In women, the superficial transverse perineal muscle is directly superior to the external sphincter, often with some overlap. In men, the superficial perineal muscle is directly anterior to the external sphincter. It closes the urogenital (levator) hiatus; supports and has a sphincter-like effect at the distal vagina; and, because it is attached to periurethral striated muscles, contributes to continence [3, 8, 13].

The superficial transverse perinei is innervated by the perineal branch of the pudendal nerve.

3.4 Pelvic Floor Attachments

The pelvic floor is directly or indirectly attached to bony structures. Direct attachments include the anterior part of the pubococcygeus muscle, including the puborectalis, which is attached to the periosteum of the posterior surface of the pubic bone. The tip of the ischial spine is the origin of the coccygeus muscle, which inserts into the lateral aspect of the coccyx and the lowest part of the sacrum. The pelvic sidewall is an important attachment site, with the internal obturator muscle as the primary constituent. This muscle has its origin at the obturator membrane, the margins of the obturator foramen, and the pelvic surfaces of the ilium and ischium. This fanlike muscle inserts at the greater trochanter of the femur. The piriformis is a posteriorly situated flat triangular shaped muscle lying directly above the pelvic floor. The internal obturator muscle and piriformis are covered by a strong membrane which has firm attachments to the periosteum. A tendinous ridge of the obturator fascia, the arcus tendinous levator ani, forms the pelvic sidewall attachment for the levator ani. This arcus extends from the pubic ramus anteriorly to the ischial spine posteriorly and most of the levator ani muscle arises from this arcus. This arcus can be identified as the junction of the levator plate and internal obturator muscle. A second tendinous arcus is the arcus tendinous fascia pelvis, whose posterior part is fused with the arcus tendinous levator ani while its anterior part has a more inferior and medial course than the arcus tendinous levator ani attaching to the pubis close to the pubic symphysis [1, 9, 11, 14]. Both arcs are linear, dense, pure connective tissue structures at the pelvic sidewall and form the attachment for several structures: the levator ani muscle, endopelvic fascia (anterior vaginal wall), pubovesical muscle, and other supportive structures.

3.5 Perineal Body

The perineal body is a pyramidal fibromuscular structure in the midline between the anus and vagina with the rectovaginal septum at its cephalad apex. Below this, muscles and their fascia converge and interlace through the structure. An important perineal structure is the perineal body (also named the *central perineal tendon*) which is a site of attachment of rectum, vaginal slips from the pubococcygeus, perineal muscles, and anal sphincter. This fibromuscular node is located at the midline between the urogenital region and the anal sphincter; it also contains smooth muscle, elastic fibers, and nerve endings. It is an important part of the pelvic floor; just above it are the vagina and the uterus [2, 17].

3.6 Rectum

3.6.1 Structure and Topographical Relations of the Rectum

The rectum is the direct continuation of the sigmoid colon and ends with the anus (Fig. 3.4). The rectosigmoid junction is approximately 6 cm below the level of the sacral promontory. The rectum in the adult measures 10–14 cm in length. The rectum, being situated above the level of the pelvic floor and below the level of the pelvic brim, is an entirely intrapelvic viscus. The first portion of the rectum is dilated and it is also known as *rectal ampulla*. The longitudinal orientation of the rectum conforms to the ventral concavity of the sacrum. Thus the rectum runs downwards and backwards initially, and then downwards and forwards to reach the levator hiatus (the gap in the pelvic floor between the two levator ani muscles through which the pelvic viscera pass inferiorly into the perineum). This natural ventral bend in the rectum is termed the sacral flexure. At the levator hiatus the rectum becomes continuous with the *anal canal* [7, 10, 12, 18].

The *anorectal junction* is situated approximately 4 cm anterior to the tip of the coccyx. In addition to presenting the ventral bend, the rectum possesses a succession of three, smooth, laterally facing curves. The upper and lower curves are directed to the right and the middle curve to the left. Each of the three "curves" presents on the inside a transverse, sickle-shaped fold. Also known as rectal shelves or the "*valves of Houston*," these folds are produced by the thickened muscle in the rectal wall projecting inwards with overlying mucosa. The middle rectal shelf is the most constant and prominent of the three shelves.

Posteriorly, the rectum is related to (i) the ventral surface of the lower half of the *sacrum* and the adjoining *coccyx*, and (ii) the *presacral space* containing the **median sacral vessels** and the **presacral venous plexus**.

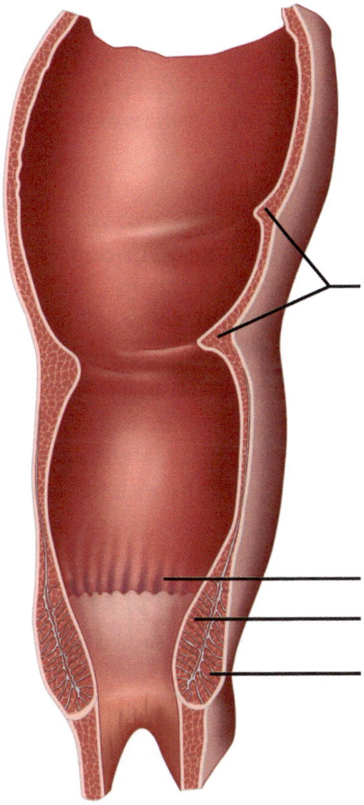

Fig. 3.4 Rectum, frontal view This image is in the public domain because it is a mere mechanical scan or photocopy of a public domain original, or—from the available evidence—is so similar to such a scan or photocopy that no copyright protection can be expected to arise. The original itself is in the public domain for the following reason: 1918, Philadelphia and New York; Publisher: Lea and Febiger
Henry Gray: Gray's Anatomy, The Anatomy of Human Body (20th edition), page 1183 https://upload.wikimedia.org/wikipedia/commons/0/01/Gray1078.png

Lumbosacral nerves and fibers from *piriformis muscle* run *posterolateral* to the rectum on either side. Both are covered anteriorly by a dense layer of pelvic fascia.

Posteroinferior to the rectum is the **anococcygeal raphe** which originates from the interdigitation of the **two levator ani muscles** (from the tip of the coccyx to the posterior aspect of the anorectal junction).

Lateral to the rectum on either side lies a condensation of pelvic fascia called the **lateral ligament** which extends from the lateral wall of the pelvis to the rectum. Just laterally to the rectum on either side, a sacral nerve, belonging to the **inferior hypogastric plexus**, runs longitudinally. It is a mixed autonomic nerve plexus, carrying which is enclosed in the *rectal fascia propria.*

Anteriorly, topographical relations of the rectum differ in the two sexes. In *males* the rectum below the peritoneal reflection is related to the posterior surface of the **urinary bladder**, the posterior surface of **seminal vesicles**, the inferior parts of the two **ureters**, and **vas**, and below the **bladder neck** and the posterior surface of the **prostate**. All of the above-named urogenital structures are separated from the fascia propria of the rectum by a distinct and fairly strong fascial layer known synonymously as rectovesical fascia, *rectovesical septum*, and *Denonvilliers' fascia* [10, 18–20].

In the female, the rectum below the level of the peritoneal reflection is related to the posterior wall of the intrapelvic **vagina**. Above the peritoneal reflection the rectum is related to the recto-uterine pouch also known as the **pouch of Douglas**. Above the vaginal fornix the rectum is related to the posterior surface of the body of the anteverted uterus [7].

3.6.2 Relationship of the Peritoneum to the Rectum

It is possible to distinguish four different surfaces of the rectum, anterior, posterior, and laterals. The upper third of the rectum is covered by peritoneum on its anterior and lateral surfaces; the middle third of the rectum is covered by peritoneum only on its anterior surface while the lower third of the rectum is below the level of the peritoneal reflection (the level at which the peritoneum leaves the anterior rectal wall to reach the viscus in front). Anteriorly to the rectum, the peritoneum covers the uterus in females and vescica in males defining the rectal-uterine space and rectovesical space *(Douglas space)* [7, 10, 21].

3.6.3 Fascial Coverings of the Rectum

The rectum is covered by a circumferential fascial layer called the *fascia propria.* This fascia, known as *mesorectum (or investing fascia of rectum)*, encloses a thin layer of perirectal fact which contains lymph nodes and vessels. The *mesorectum* is not a proper suspensory mesentery. The entire length of the rectum (except perhaps the very distal centimeter) is surrounded by a cuff of fat termed the *perirectal fat*, which is generally more abundant posteriorly than anteriorly. It is in this perirectal fat that the epirectal and pararectal lymph nodes are located, and it is in this perirectal fat that the superior rectal vessels travel before entering the rectum. Anteriorly there is a strong fascial layer known as *Denonvilliers' fascia* that separates the rectum from the urinary system in males [7, 10, 21].

3.6.4 Arterial Supply and Venous Drainage of the Rectum

The principal artery supplying the rectum is the **superior rectal artery**, which is the terminal branch of the *inferior mesenteric artery* at the point where it crosses the pelvic brim (Fig. 3.5). The *superior rectal artery* runs with the pelvic

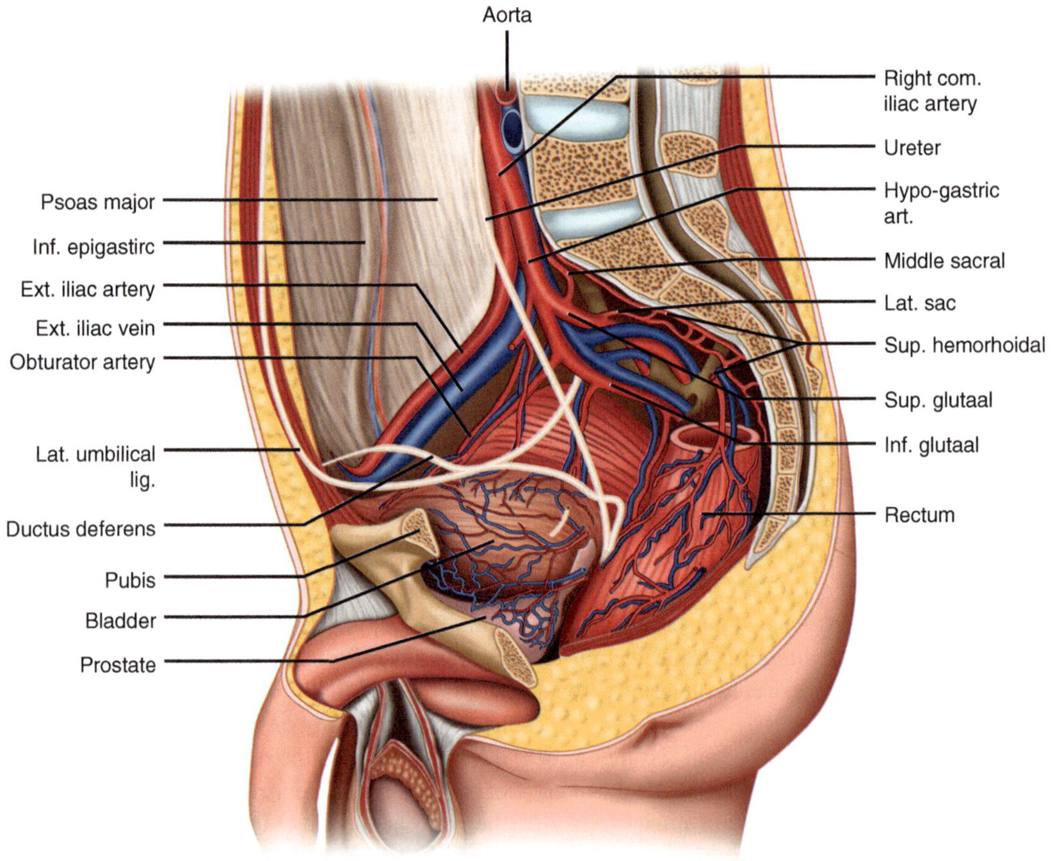

Fig. 3.5 Pelvic vascularization This image is in the public domain because it is a mere mechanical scan or photocopy of a public domain original, or—from the available evidence—is so similar to such a scan or photocopy that no copyright protection can be expected to arise. The original itself is in the public domain for the following reason: 1918, Philadelphia and New York; Publisher: Lea and Febiger **Henry Gray**: Gray's Anatomy, The Anatomy of Human Body (20th edition), page 613 https://upload.wikimedia.org/wikipedia/commons/3/3e/Gray539.png

attachment of the sigmoid mesocolon to enter the perirectal fat behind the rectum. At this point the artery gives 2 or 3 branches which travel longitudinally on each side of the rectum. Supplementary arteries which contribute to the blood supply of the rectum are the **middle rectal arteries**, the **inferior rectal arteries**, and the **median sacral artery**. The *right and left middle rectal arteries* arise from the corresponding **internal iliac artery** and run inferomedially just above the pelvic floor to reach the rectum. The middle rectal arteries are inconstant in size and are usually not prominent vessels. They may be enclosed within the *lateral ligament*. Each *inferior rectal artery* is a branch of the **internal pudendal artery** and is given off as soon as the latter enters the perineum. The *inferior rectal artery* crosses the ischio-anal fossa from lateral to medial to enter the anal wall. The **median sacral artery** arises from the posterior aspect of the aorta just proximal to the aortic bifurcation. It runs down the anterior aspect of the sacrum and on reaching the pelvic floor it runs anteriorly to terminate in the rectal wall [10, 12, 19].

The *venous drainage* of the rectum mirrors the arterial supply. From a rich and valveless intramural venous plexus blood enters the valveless, perirectal venous plexus, whence rectal blood is carried mainly in the superior rectal vein. The superior rectal vein running alongside the artery crosses the pelvic brim from below upwards to become the inferior mesenteric vein. Thereafter the **inferior mesenteric vein** drains the sigmoid colon, descending colon, and splenic flexure before emptying into the splenic vein and thereby into the portal vein. Some venous blood from the intramural and perirectal venous plexuses travels bilaterally in the middle rectal veins and drains into the **internal iliac veins**. Venous blood from these rectal plexuses also finds its way through the anal wall into the inferior rectal veins which drain into the internal iliac veins via the internal pudendal veins [7, 10, 19].

The *anal mucosa and submucosa* thus represent sites of natural **portosystemic venous anastomoses**. To a limited extent these anastomoses are also present in the rectal wall.

3.6.5 Lymphatic Drainage of the Rectum

As with the lymphatic drainage of the colon, rectal lymph is initially received by the lymphoid follicles in the mucosa.

Thereafter, the lymph passes successively through three tiers of **mesorectal lymph nodes** (equivalent to *epicolic, paracolic, and intermediate nodes*) before reaching the principal nodes. The principal lymph nodes that receive most of the lymph from the upper two-thirds of the rectum are the **inferior mesenteric lymph nodes** which are situated around the origin of the inferior mesenteric artery. Lymph from the lower third of the rectum drains into three sets of principal nodes: the *inferior mesenteric lymph nodes* and the *internal iliac lymph nodes bilaterally* (also called the pelvic sidewall nodes) [10].

3.6.6 Innervation of the Rectum

The nerve supply of the rectum is by the autonomic system. Sympathetic supply is by branches of the superior hypogastric plexus and by fibers accompanying the inferior mesenteric and superior rectal arteries from the celiac plexus. Parasympathetic (motor) supply is from S2–S4 to the inferior hypogastric plexuses by the pelvic splanchnic nerves (nervi erigentes). These fibers give sensory supply (crude sensation and pain) and have a role in discriminating between flatus and feces [7].

3.7 Anal Canal and Sphincter

3.7.1 Structure and Topographical Relations of the Anal Canal

The **anal canal** is the very terminal segment of the alimentary tract, and lies entirely below the level of the pelvic floor in the region termed the perineum. The anal canal ends at the anal verge, forming a transitional zone between the epithelium and the perianal skin [10, 18].

The anatomic anal canal is defined as beginning at the dentate line and ending at the anal verge. On the other hand, the surgical anal canal is defined as the area between the anorectal ring and the anal verge *(the internal anal sphincter, external anal sphincter, and puborectalis muscle constitute the surgical anal canal)*. The anal canal in the adult is about 4–5 cm in length, while it is still debated in children. From its commencement in the **levator hiatus** as the direct continuation of the *rectum*, the *anal canal* passes downwards and backwards. The consequent bend at the *anorectal junction* (acute angle directed posteriorly) is termed the perineal flexure, and is produced by the forward pull of the sling-like puborectalis muscle, which is of course derived from both levator ani muscles [7, 12, 22].

The anal canal has a cylindrical double-layered shape. The inner layer is composed of the internal anal sphincter and the conjoined longitudinal muscle (CLM), which is innervated by the autonomic nerve system, whereas the outer layer is composed of the puborectalis muscle and the external sphincter muscle (EAS), which is stimulated by somatic nerves. The EAS finishes more distally than the internal sphincter. The EAS constitutes the end of the anal canal, just as it surrounds the IAS from the outside to the inside [6, 7].

3.8 Anal Mucosa Lining

The upper half of the anal canal (i.e., above the dentate line) presents a variable number of vertical mucosal ridges (8–10) termed anal columns or the columns of Morgagni. Running between the lower ends of adjacent columns are a series of curved folds of mucosa called anal valves.

It is these anal valves circumferentially arranged that account for the dentate/pectinate line. Above each anal valve is a shallow mucosal pocket termed the anal sinus. Opening into each anal sinus are the ductules of a number of mucus-secreting anal glands. These glands are situated in the anal mucosa, submucosa, and even as deeply as the internal sphincter of the anal canal. The epithelium above the dentate line is similar to the glandular epithelial lining of the rectal mucosa and is made up of columnar cells, crypts, and goblet cells. About 10 mm proximal to the dentate line, the "anal transitional zone" is found. It appears purple and represents an area of gradual transition from a squamous epithelium to a columnar epithelium. The anal canal distal to the pectinate line is, by sharp contrast, lined with non-keratinized, stratified squamous epithelium, and presents a smooth appearance. Distally still, at the anal verge and just proximal to it, the anal canal is lined with sensitive, thick, hair-bearing skin.

3.8.1 Arterial Supply and Venous Drainage of the Anal Canal

The *arterial supply* of the **external** and **internal anal sphincters** as well as the mucosa over the lower half of the anal canal is derived from the *right and left inferior rectal arteries*. The mucosa proximal to the dentate line is, however, supplied by terminal twigs of the *superior rectal artery*. The middle rectal arteries make an insignificant contribution. Within the wall of the anal canal there is a rich anastomosis between the terminations of the inferior rectal and *superior rectal arteries* and, for what it is worth, the terminal twigs of the middle rectal arteries.

The *veins draining* the anal canal correspond to the main arteries that supply the anal canal, and originate in the venous plexus situated in the anal wall which is continuous with the intramural rectal venous plexus. Venous channels from the upper part of the anal canal (proximal to the dentate line) drain mainly into the **superior rectal vein** and thereby eventually to the **portal venous system**. Distal to the dentate line, venous drainage is mainly to the **internal iliac veins** either directly via the **middle rectal veins** or indirectly via the **inferior rectal veins** and **internal pudendal veins**. In addition to the intramural venous plexuses in the anal wall, and very possibly related to them, are arteriovenous mucosal cushions situated in the upper half of the anal canal [10, 12, 22].

3.9 Internal Sphincter

The **internal anal sphincter (IAS)** is approximately 3–4 cm long and is the distal extension of the *inner circular muscle* layer in the rectal wall. It is a specialized entity and may be up to 5 or 6 mm in thickness. Being made up of *smooth involuntary muscle*, its motor innervation is derived from the autonomic nerves, sympathetic (spinal nerves) and parasympathetic (pelvic nerves) [23]. The distal edge of the internal sphincter is well demarcated and in the normal individual it is usually easily felt.

The IAS contributes 60–75% of the resting anal tone, while the external anal sphincter and puborectalis collectively contribute about 20%. The remainder is provided by the dilated anal mucosal cushions which have already been referred to above [24].

3.10 Conjoined Longitudinal Muscle

The CLM, also called the longitudinal anal muscle, has been described as a vertical layer of muscular tissue interposed between the IAS and the EAS. This muscle lies on the intersphincteric space, which contains a thin layer of fatty tissue. The CLM begins at the anorectal ring as an extension of the longitudinal rectal muscle fibers and descends caudally [25]. Measuring approximately 2.5 mm in thickness, the CLM decreases in thickness with age. At the most caudal part of the CLM, smooth muscles called the "corrugator cutis ani" traverse the distal anal sphincter into the perianal skin and the ischiorectal fossa. Those smooth muscles also pass through the IAS to intersect within the subepithelial space.

3.11 External Anal Sphincter

In his original description Santorini (1769) stated that EAS has three separate muscle bundles: subcutaneous, superficial, and deep. However, several investigators have found that the *subcutaneous* and *superficial* muscle bundles only constitute the EAS. The **subcutaneous** portion of the EAS is located caudal to the IAS and the superficial portion surrounds the distal part of IAS. The **deep** portion of the EAS is either very small or merges imperceptibly with the *puborectalis muscle* (Fig. 3.6).

The EAS consists of a striated muscle that envelops the entire IAS and CLM. The EAS is approximately 2.7 cm high. The average thickness of the EAS is 8.6 mm in males and 7.7 mm in females, respectively. The anterior part of EAS differs between the sexes. In women, the muscle fibers unite in the inferior part anteriorly and so imaging just above this level may show an apparent defect. The posterior wall of the EAS is shorter in its cranio-caudal extent than the anterior wall. In men the EAS is symmetrical at all levels. Anteriorly, the EAS is attached to the perineal body and transverse perinei muscle, and posteriorly to the anococcygeal raphe.

The muscle fibers of EAS are composed of *fast*- and *slow*-twitch skeletal muscle types, which allow it to maintain sustained tonic contraction at rest (its contribution is estimated to be around 25%) and also allow it to contract rapidly with voluntary squeeze. Motor neurons in *Onuf's nucleus* (located in the sacral spinal cord) innervate EAS muscle through the inferior rectal branches of the right and left pudendal nerves (S2, S3) and the perineal branch of the fourth sacral nerve (S4) [1, 22].

3.11.1 Functional anatomy

3.11.1.1 Anal Continence

The anal continence is dependent on four structures: the *IAS*, the *EAS*, the *puborectalis sling* (the latter being derived from the levator ani muscles of the two sides), and the *arteriovenous mucosal cushions.*

The **IAS** is approximately 3–4 cm long and is the distal extension of the inner circular muscle layer in the rectal wall. It is nevertheless a specialized entity and may be up to 5 or 6 mm in thickness. Being made up of smooth *involuntary muscle*, its motor innervation is derived from the autonomic nervous system (predominantly sym-

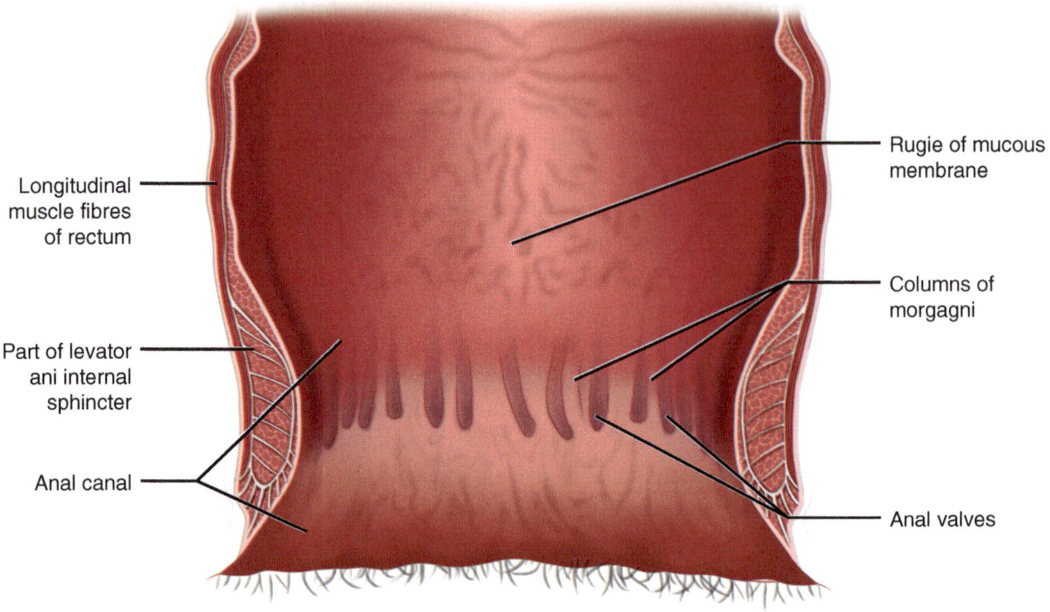

Fig. 3.6 Anal canal This image is in the public domain because it is a mere mechanical scan or photocopy of a public domain original, or—from the available evidence—is so similar to such a scan or photocopy that no copyright protection can be expected to arise. The original itself is in the public domain for the following reason: 1918, Philadelphia and New York; Publisher: Lea and Febiger
Henry Gray: Gray's Anatomy, The Anatomy of Human Body (20th edition), page 1184 https://upload.wikimedia.org/wikipedia/commons/9/9f/Gray1080.png

pathetic). The IAS contributes 60–75% of the resting anal tone, while the **EAS** and **puborectalis** collectively contribute about 20%. The remainder is provided by the dilated **anal mucosal cushions.** The EAS is longer and wider than the IAS, and the distal edge of the EAS is normally distal to that of the internal sphincter by at least 1 cm. Between these two edges, it is relatively simple to palpate the intersphincteric groove.

The **puborectalis sling** comprises those fibers of each levator ani muscle which arise from the periosteum on the posterior surface of the pubic bone 1 cm, or more, lateral to the pubic symphysis. These fibers run posteriorly and swing medially behind the recto-anal junction to meet their counterparts from the other side in a "U"-shaped muscle. Together these fibers form a sling behind the recto-anal junction. Contraction of the puborectalis muscle lifts up the anal canal in the ventral or anterior direction and in so doing causes compression of the anal canal, vagina, and urethra against the back of pubic symphysis. The constant tonic contraction in this sling accounts for the sharp recto-anal angle. Voluntary relaxation of the puborectalis sling allows straightening of the recto-anal tube, a prerequisite to defecation. The puborectalis muscle is innervated like the rest of levator ani by the ipsilateral perineal branch of S4, a branch of the lumbosacral plexus. The deepest part of the EAS blends with the puborectalis sling behind the anorectal junction. This area of fusion is palpable on per rectal digital examination, and in surgical terminology is referred to as the anorectal ring [7, 22, 26–28].

References

1. Raizada V, Mittal RK. Pelvic floor anatomy and applied physiology. Gastroenterol Clin N Am. 2008;37:493–vii.
2. Herschorn S. Female pelvic floor anatomy: the pelvic floor, supporting structures, and pelvic organs. Rev Urol. 2004;6:S2–10.

3. Ashton-Miller JA, DeLancey JOL. Functional anatomy of the female pelvic floor. Ann N Y Acad Sci. 2007;1101:266–96.
4. Strohbehn K. Normal pelvic floor anatomy. Obstet Gynecol Clin N Am. 1998;25:683–705.
5. Eickmeyer SM. Anatomy and physiology of the pelvic floor. Phys Med Rehabil Clin N Am. 2017;28:455–60.
6. Macchi V, Porzionato A, Stecco C, Vigato E, Parenti A, De Caro R. Histo-topographic study of the longitudinal anal muscle. Clin Anat N Y N. 2008;21:447–52.
7. Stoker J. Anorectal and pelvic floor anatomy. Best Pract Res Clin Gastroenterol. 2009;23:463–75.
8. Bharucha AE. Pelvic floor: anatomy and function. Neurogastroenterol Motil Off J Eur Gastrointest Motil Soc. 2006;18:507–19.
9. Lawson JO, Pelvic anatomy. I. Pelvic floor muscles. Ann R Coll Surg Engl. 1974;54:244–52.
10. Mahadevan V. Anatomy of the rectum and anal canal. Surg Oxf Int Ed Elsevier. 2020;38:7–11.
11. Raychaudhuri B, Cahill D. Pelvic fasciae in urology. Ann R Coll Surg Engl. 2008;90:633–7.
12. Lee JM, Kim NK. Essential anatomy of the anorectum for colorectal surgeons focused on the gross anatomy and histologic findings. Ann Coloproctol Korean Soc Coloproctol. 2018;34:59.
13. DeLancey JO. Structural support of the urethra as it relates to stress urinary incontinence: the hammock hypothesis. Am J Obstet Gynecol. 1994;170:1713–20. discussion 1720–1723
14. Wei JT, De Lancey JO. Functional anatomy of the pelvic floor and lower urinary tract. Clin Obstet Gynecol LWW. 2004;47:3–17.
15. Grigorescu BA, Lazarou G, Olson TR, Downie SA, Powers K, Greston WM, et al. Innervation of the levator ani muscles: description of the nerve branches to the pubococcygeus, iliococcygeus, and puborectalis muscles. Int Urogynecol J Pelvic Floor Dysfunct. 2008;19:107–16.
16. Niikura H, Jin ZW, Cho BH, Murakami G, Yaegashi N, Lee JK, et al. Human fetal anatomy of the coccygeal attachments of the levator ani muscle. Clin Anat. 2010;23:566–74.
17. Klutke CG, Siegel CL. Functional female pelvic anatomy. Urol Clin North Am. 1995;22:487–98.
18. Church JM, Raudkivi PJ, Hill GL. The surgical anatomy of the rectum—a review with particular relevance to the hazards of rectal mobilisation. Int J Color Dis. 1987;2:158–66.
19. Heald RJ, Moran BJ. Embryology and anatomy of the rectum. Semin Surg Oncol. 1998;15:66–71.
20. Kourambas J, Angus DG, Hosking P, Chou ST. A histological study of Denonvilliers' fascia and its relationship to the neurovascular bundle. Br J Urol. 1998;82:408–10.
21. Chang Y, Liu HL, Jiang HH, Li AJ, Wang WC, Peng J, et al. Aatomical relationship between fascia propria of the rectum and visceral pelvic fascia in the view of continuity of fasciae. Zhonghua Wei Chang Wai Ke Za Zhi Chin J Gastrointest Surg. 2019;22:949–54.
22. Lunniss PJ, Phillips RKS. Anatomy and function of the anal longitudinal muscle. Br J Surg Wiley Online Library. 1992;79:882–4.
23. Bharucha AE, Klingele CJ. Autonomic and somatic systems to the anorectum and pelvic floor. Peripher Neuropathy Elsevier Inc. 2005:279–98.
24. Rattan S. The internal anal sphincter: regulation of smooth muscle tone and relaxation. Neurogastroenterol Motil Wiley Online Library. 2005;17:50–9.
25. Haas PA, Fox TA. The importance of the perianal connective tissue in the surgical anatomy and function of the anus. Dis Colon Rectum Springer. 1977;20:303–13.
26. Gibbons CP, Bannister JJ, Trowbridge EA, Read NW. Role of anal cushions in maintaining continence. Lancet Elsevier. 1986;327:886–8.
27. Lierse W, Holschneider AM, Steinfeld J. The relative proportions of type I and type II muscle fibers in the external sphincter ani muscle at different ages and stages of development-observations on the development of continence. Eur J Pediatr Surg\copyright Georg Thieme Verlag KG Stuttgart New York. 1993;3:28–32.
28. Keef KD, Cobine CA. Control of motility in the internal anal sphincter. J Neurogastroenterol Motil. 2019;25:189–204.

Physiology of the Pelvic Floor and the Anorectum

4

Eleni Athanasakos and Stewart Cleeve

4.1 Introduction

It is extraordinary how early on during gestation that the primitive gut tube develops (during 3–4 weeks) by incorporating the yolk sac during craniocaudal and lateral folding of the embryo [1]. The anatomy of the colon, rectum and anus originates from the nineteenth and early twentieth centuries, with a detailed appreciation made as early as 1543 by Andreas Vesalius through anatomical dissections [2]. The anatomy cannot be solely appreciated without the physiological processes involved, which are impacted by disease or dysfunction.

The central role of the colon is to serve as a temporary storage area for the body waste and preserve water and energy [3]. The colon needs to be able to propel intestinal content distally while simultaneously reabsorbing liquid from that content. A prolonged transit is needed to allow definite homeostatic functions [4] including (1) the right side of the colon (i.e. caecum and ascending colon) which is responsible for absorption of water, nutrients and electrolytes and fermentation of undigested sugars and (2) the left side of the colon which includes the storage and controlled evacuation of faecal material [5]. The primary functions of the rectum include (1) the temporary storage of faeces prior to the process of evacuation, (2) synchronisation of the anal canal to recognise the urge to defaecate and iii) propulsion of contents towards the anal canal during defaecation. The interaction of anal sphincters and other associated muscles of the pelvic floor and rectum are the key players for the preservation of continence [6].

This chapter discusses the physiological mechanisms of continence and defaecation.

4.2 Mechanism of Continence

Continence is maintained by structural and functional integrity and compliance of the rectum [7]. This is orchestrated by the rectum, anus and associated pelvic floor musculature; rectal compliance; capacity and sensations; colonic transit; stool consistency; and central nervous system and somatic and visceral input [8–10]. This section will individually discuss the physiology of continence and how it is controlled and maintained in humans.

E. Athanasakos (✉) · S. Cleeve
Department of Paediatric Surgery, Level 8, Paediatric Offices, The Royal London Hospital, Barts Health, Whitechapel, UK
e-mail: eleni.athanasakos@nhs.net; stewartjames.cleeve@nhs.net

M. Lima, G. Ruggeri (eds.), *Ano-Rectal Endosonography and Manometry in Paediatrics*,
https://doi.org/10.1007/978-3-030-97668-2_4

4.2.1 Anal Sphincter Muscles

There are two sophisticated anal sphincters which perform a significant role in maintaining faecal continence. At rest, the anal canal remains closed to preserve continence, which is by the tonic activity of the internal (IAS) and external anal sphincter (EAS), together with the anal cushions [11]. The sphincteric complex is rather dynamic and remarkable, as it is controlled by a variety of reflexes and modulation by higher centres, allowing the anal canal to be sealed at all times except when we want to pass flatus or defaecate [12]. The primary role of the IAS is to keep us clean which is represented by our unconscious inner world. The EAS on the other hand is our servant of our consciousness, thus allowing us to defaecate only when suitable and remaining closed. These two sphincter muscles have to work as a team; however, we will separately explain in detail their role.

4.2.1.1 Internal Anal Sphincter (IAS)

As discussed in the previous chapter, the inner circular muscle layer of the rectum expands caudally into the anal canal to become the IAS. Within the sphincteric complex, the circular muscles are thicker than those of the rectal circular smooth muscle with discrete septa in between the muscle bundles [13].

It is the activity of the anal sphincters and the puborectalis muscle which produces the zone of high pressure inside the anal canal [8, 14]. Since the IAS is innervated by the autonomic nervous system, it is not subject to voluntary control. Thus, this muscle exists in a continuously tonic state, is essential for maintaining the closure of the resting pressure of the anal canal and initiates the act of defaecation by reflex dilation in response to rectal distension [7–9, 14].

The IAS has been confirmed electromyographically to demonstrate a constant activity at rest to preserve continence [15–17]. Remarkably, it is unaffected by respiration, state of wakefulness or feeding or administration of a general anaesthesia [18], not even by the activity of its neighbouring EAS [15, 19, 20]. Circadian differences have been demonstrated to be dependent on the sleep–wake cycle and ultradian (20–40 min in length) rhythms (i.e. where cycles are repeated) are independent of the sleep–wake cycle [21].

IAS has resting tonus with cyclic variations by the short and ultrashort waves known as 'slow-twitch' or 'slow-wave', fatigue-resistant smooth muscles with a frequency ranging from 10 to 40 cycles per minute [15, 18–20, 22, 23]. Ultrashort waves have been found in a wide range of 5-90% of individuals, with frequencies ranging from 0.5 to 2 cycles per minute and which has been demonstrated to decrease during sleep [19, 23, 24]. The significance of these waves is in reference to a reflex of accommodation or sampling reflex in the upper part of the IAS, which will be discussed in more detail later. Approximately, with a frequency of seven times an hour, the IAS intermittently relaxes for 10–20 s [25, 26]. This intermittent relaxation period allows for equilibrium between the pressure of the anal canal and rectum, permitting contact to the rectal contents with the sensitive mucosa of the anal canal.

The IAS contributes approximately 85% of the anal resting pressure (i.e. measured at between 50 and 120 mmHg in a healthy adult) [6, 27, 28] and normally around 30–40 mmHg in children [29]. This adequate sphincter anal pressure allows continence, unless the sphincter is inhibited via the myenteric plexus and the recto-anal reflex (RAIR), when the rectum is distended or contracted [29]. However, there has been debate regarding the influence of IAS activity being 85% at rest, 65% during constant rectal distension and 40% after sudden rectal distension [22], while another study estimating lesser influence, in that approximately 55% of resting anal tone is due to IAS activity [30]. Approximately 15–20% of the anal resting tone is represented also by the puborectalis, EAS muscles, vascular cushions and mucosal folds [26, 31], which are controlled by the intrinsic myogenic activity and extrinsic adrenergic innervation. Although we know that the resting sphincter tone is predominantly attributed to the IAS, studies have demonstrated under general anaesthesia [32] or after pudendal nerve block [33–35] that there is a tonic excitatory sympathetic discharge to IAS and that this excitatory sympathetic discharge does not contribute to anal pressure during rectal distension [36].

The IAS is supplied by the autonomic nerves including (1) sympathetic (spinal nerves) which originate from the lower thoracic ganglia forming the superior hypogastric plexus and (2) parasympathetic (pelvic nerves) which originate from the 2nd–4th sacral nerves forming the inferior hypogastric plexus, which gives rise to superior, middle and inferior rectal nerves, supplying the rectum and anal canal [13, 37].

The pharmacological regions found in the IAS can be divided accordingly: (1) proximal part of the IAS where acetylcholine causes contraction (i.e. as in the rectum and rest of the alimentary tract) and nitric oxide (NO) will cause relaxation, where ganglion cells are found and decrease in numbers as it reaches the anal direction, and (2) more distally, alpha-stimulating, beta-relaxing receptors and noradrenergic noncholinergic (NANC) relaxing nerve fibres found around [38]. Most of the IAS tone is myogenic due to the smooth muscle nature, with angiotensin II and prostaglandin F2α contributing modulatory roles. IAS contraction is mediated by the sympathetic nerves, which are stimulated by α and relaxation of the IAS through β1, β2 and β3 adrenergic receptors [13]. IAS relaxation is stimulated by the parasympathetic or pelvic nerves, through nitric oxide (NO)-containing neurons located in the myenteric plexus [39] and Cajal cells [13, 38] with other inhibitory neurotransmitters found (e.g. vasointestinal intestinal peptide (VIP) and carbon monoxide) but with less contribution. Degeneration of myenteric neurons causing impaired IAS relaxation is the trademark of Hirschsprung's disease [40].

Several studies have shown that weakness or disruption of the IAS results in the passive faecal and/or flatus incontinence [41] which are related to low resting anal tone due to impairment of the IAS [41–44]. Rupture (i.e. partial or more) has been demonstrated in surgical procedures such as dilatations which were formerly used to treat anal fissures or use of partial sphincterotomy [44] where division or rupture of the sphincteric complex leads to impairment of the continence mechanisms. Furthermore, low anterior rectal resection and coloanal anastomosis can also lead to physiological impairment of the IAS and neighbouring muscles [26, 43].

4.2.1.2 External Anal Sphincter (EAS)

The EAS is a striated muscle which surrounds the IAS and extends down to the skin at the anal verge and innervated by the pudendal nerve. There have been several studies dating back from Santorini (1769) [45] to present, discussing the anatomy of the EAS about its structure and subdivisions or if it is part of other associated musculature [46–51]. However, this chapter will not focus on the anatomical discussion of the EAS as in previous chapters, but more regarding its role.

As previously mentioned, both the involuntary IAS and voluntary EAS are team players which play an important role in maintaining continence and both have a resting tone. The EAS contributes a small amount to anal resting pressure, but is responsible for the squeeze pressure. Without this high resting pressure in the anal canal, we would be unable as humans to prevent leakage of mucus and gas [8, 52]. In fact, the EAS is in a state of constant tonic activity at rest [53], even during sleep [53–56], and makes up about 30% of the basal resting anal tone [30]. This is an unusual nature of a striated muscle that opportunely persists during sleep.

The EAS is composed of both tonically contracting 'slow-twitch' fibres and phasically contracting 'fast-twitch' fibres [6]. Unlike the IAS, the EAS is a fatigable muscle; when maximally contracted, for up to 50 s to 1 min in adults [6, 29, 57, 58], fatigue occurs, and about 30 s in children [29]. This is of clinical relevance, when considering whether constipation due to deliberate withholding of stools is possible in every child or just those with a physical predisposition [29]. The EAS contracts for approximately 20–30 s, when the rectum distends from the arrival of gases or faeces in the rectal ampulla—this is known as 'guarding reflex' by a low spinal reflex with cortical control that can be found in patients who have had coloanal anastomosis, once it is triggered by receptors located in the puborectalis region [59, 60]. As previously mentioned, the anal canal is usually closed at rest and during sleep, as reinforced by the constant activity of the IAS and the tonic activity of the EAS and puborectalis muscle.

Continence is preserved by the contraction of the EAS to enhance the anal tone, when continence is threatened. When the EAS contracts, this squeeze response may be either voluntary or induced by increased intra-abdominal pressure [53] or simply moving your finger across the anal canal lining (e.g. during a PR exam) [57]. The voluntary contraction of the EAS can range with pressures between 50 and 200 mmHg being generated and obstetric EAS injury is associated with a significant reduction in maximum voluntary squeeze pressure [61]. In addition to the EAS, at rest, the levator ani muscles (LAM) and the puborectalis also remain in a state of continuous contraction by the reflex known as the postural reflex which is maintained through the lower lumbar and sacral spinal cord and provides assistance to and supports the weight of the pelvic viscera [56].

Poor functional outcome may result from anal sphincter dysfunction, either as a result of organ dysgenesis or from iatrogenic or obstetric injury. The integrity of both the IAS and EAS is important for a good functional result following surgical repair for anorectal malformations within paediatrics [62–65]. The integrity importance of these anal sphincters can also be demonstrated when faecal or flatus incontinence is found in up to 50% of women who present with previous obstetric injury and defect [66].

4.2.2 Anorectal Angle

The word rectum originates from the Latin 'rectum' to mean 'straight' or 'regular' as it appears in mammals. Yet, in humans the rectum actually has curves which adapt to the anatomical shape of the sacrum and coccyx [67]. It is well known that the rectum is not a straight tube—posteriorly it curves above the puborectalis sling (i.e. the anorectal angle), following the concave anterior surface of the sacrum towards the promontory, interlinked towards the direction of three functional folds (i.e. valves of Houston), before it angles sharply as it enters the pelvis (i.e. rectosigmoid angle) [68]. It has been widely recognised for years that the anorectal angle is the most effective mechanism of valvular control for faecal continence [69, 70].

The angle is formed by the crossing of the upper limit of the anal canal with the median line of the rectum, which is supported by the puborectalis muscle [71]. Radiological studies (MRI or barium defecography) have demonstrated the location of the anorectal angle in relationship with the pubococcygeus line [72, 73] with the latter being an imaginary line connecting the lower edge of pubic symphysis and the tip of the coccyx. Bharucha et al. (2005) [74] have demonstrated using MR imaging how the anorectal angle changes during various movements such as when we squeeze and during defaecation [74, 75]. These changes are determined by the anorectal angle reflecting the movement of the puborectalis muscle—the constrictor function of pelvic diaphragm muscle [13]. However, it has also been related to other neighbouring muscles as well including the pubococcygeus, iliococcygeus and ischiococcygeus muscles.

At rest, it is the contractile traction of the puborectalis muscle which maintains the anorectal angle at approximately 90 degrees [6, 76]. Although the purpose of the anorectal angle assists in the preservation of continence [77] increased acuity has been associated with obstructed defaecation [78, 79]. The anorectal angle will be further discussed later, in relation to the defaecation mechanism.

4.2.3 Puborectalis and Levator Ani Muscles (LAM)

At rest, the LAM, the puborectalis and EAS remain in a state of continuous contraction. This reflex is known as the 'postural reflex' and supports the weight of the pelvic and abdominal viscera [56]. The reflex is controlled by the lower lumbar and sacral spinal cord [56].

As discussed previously, the puborectalis muscle is a component of the diaphragm, which creates a forward pull that maintains the anorectal angle and is a key player in the preservation of continence [6, 80]. The puborectalis muscle, just like the EAS, is striated and voluntarily contracts to close the anal canal, but unlike the EAS, it has the ability to sling round the upper anal canal, inserting into the pubic bone and thus increasing the anorectal angle [6]. The tonically active puborectalis muscles contract during a sudden rise in the intra-abdominal pressure, reducing the anorectal angle and thus preserving continence [80, 81]. Parks (1975) [81] suggested the 'flap-valve theory' of anal incontinence; that is, a rise in intra-abdominal pressure drives the anterior rectal wall into the upper anal canal, thereby occluding it [82]. There has been debate within the literature regarding the innervation of the puborectalis muscle, with cadaveric studies suggesting that it is supplied by the pudendal nerve. However, electrophysiological stimulation studies in humans have implied that the puborectalis muscle is in fact supplied by branches arising from the sacral plexus above the pelvic floor [82].

Disruption of the puborectalis muscle inevitably causes significant FI, underscoring the importance of this muscle in maintaining continence [83]. The puborectalis muscle can maintain continence even in the absence of anal sphincters [84, 85]. As discussed, it is evident that there are three key distinct anatomical structures, i.e. the IAS, EAS and puborectalis muscle, all of which support anal canal opening. Studies have greatly focused on the dysfunction of the EAS and IAS, both being the main culprits to FI [25, 61, 86, 87]. However, more recent studies have suggested a more 'multi-hit hypothesis'. Fernandez-Fraga et al. (2002) [87] used a novel instrument 'perineal dynamometer' to examine the function of the pelvic floor, in patients with FI. They demonstrated that these patients actually have functional defect in more than one muscle of the anal complex and the severity of FI was associated with deficit or damage to the three continence muscles.

It has been well accepted that the anal sphincters play an essential role in faecal continence, yet some patients with severe sphincter defects remain continent or experience only a mild form of incontinence, and other patients with mild sphincter defects can be severely incontinent [88, 89]. Broen et al. (2018) [88] recently demonstrated a new theory based on this, that the puborectalis may have both voluntary and involuntary contractions using anorectal function tests in 23 healthy adults. They observed during the balloon retention test involuntary contraction of the puborectal muscle, which gradually increased during progressive filling of the rectum. They called this mechanism the 'puborectal continence reflex'.

The role of the LAM is to support the pelvic floor and associated abdominal organs, thus preventing excessive perineal descent, which is mediated by a pelvic reflex known as 'postural reflex'. This reflex provides a state of constant and active contraction of the LAM, which is innervated by the sacral nerves (S2–S4) [55, 57]. LAM trauma occurs in between 15% and 35% of vaginally parous women [90]. Fernandez-Fraga et al. (2002) [87] demonstrated in incontinent women that LAM contraction was independently correlated with the severity of incontinence. However, it remains unclear, based on the anatomical proximity of the puborectalis muscle and EAS, with limited physiological tools to differentiate between the structures [6].

4.2.4 Rectoanal Inhibitory Reflex (RAIR)

The transition zone within the anal canal epithelium is lined with a large number of nerve endings and sensory cells derived from sensory, motor and autonomic nerves and enteric nervous system [91, 92]. These sensory nerve endings respond to various stimuli (e.g. pressure, temperature and friction) [93]. Distension of the rectal wall by the intestinal contents directly stimulates the pressure receptors on the wall of the rectum. This is where the mechanism of the rectoanal inhibitory reflex (RAIR) becomes important. This occurs by a brief conscious contraction of the EAS, puborectalis and pelvic floor musculature which propels the contents in an oral direction [94], allowing the IAS to recover following relaxation in response to rectal distension (Fig. 4.1) [95]. It is therefore during this 'sam-

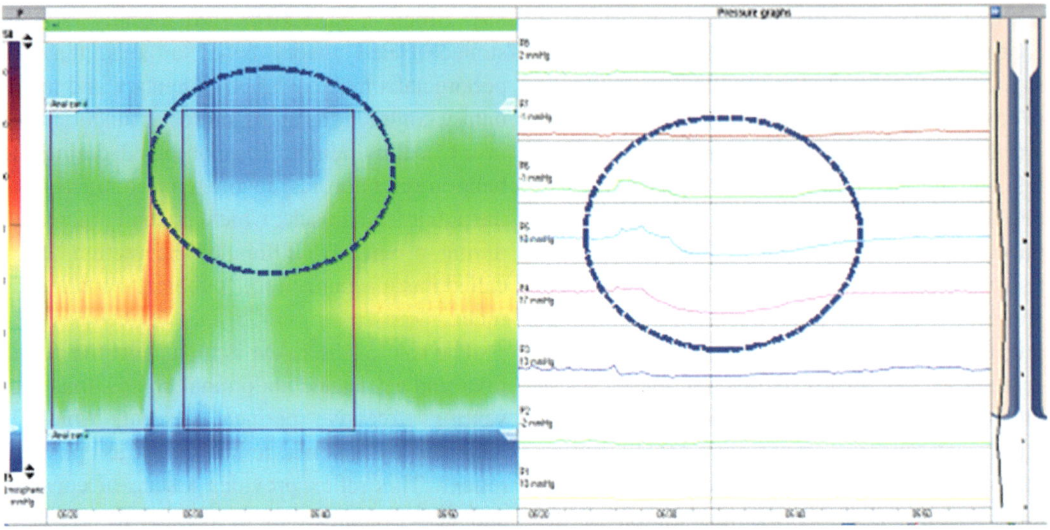

Fig. 4.1 The presence of RAIR in a child (reprinted with permission) [95]

pling' response that continence will be sustained by synchronous contraction of the EAS, which permits time for impulses to reach conscious awareness (i.e. the individual can decide how to respond) [96, 97]. At this time, the rectum can then accommodate the faecal contents, due to decrease in rectal pressures and its capacity of compliance, contributing to anal continence. The sampling reflex occurs about seven times per hour [26] (Miller and Lewis) in healthy control individuals, but less often in patients with FI [26]. The importance of the anal 'sampling' allows sensory epithelium of the anal canal the chance to differentiate between solids and liquids from flatus [98, 99].

The reflex is controlled by the enteric nervous system [100, 101], with a degree of regulation from the sacral cord [100], and is absent in patients suffering from Hirschsprung's disease (HSCR) [102]. Research has demonstrated that a RAIR can develop in children after a pull-through operation for HSCR [103–106]—particularly following transanal procedures [107]. Morikawa et al. (1987) [106] demonstrated the presence of a RAIR in 39% of patients with HSCR after the Soave Denda technique.

Furthermore, the consequences of an attenuated RAIR may lead to evacuation difficulties and equally an exaggerated reflex could result in FI [6]. Studies have shown that the RAIR can be preserved after spinal cord injury and extrinsic denervation of the rectum [32, 35, 108–110] and after rectal excision and colo/pouch anal anastomoses, with the ability to differentiate flatus from faeces [99]. However, these studies have taken into consideration the functional importance of the length of the rectal stump left in situ during rectal excision. It has been found that the smaller the rectal stump, the worse the functional outcome of continence, in relation to rectal sensation and compliance and above any changes in the RAIR [111–114]. Thus, there is limited evidence to suggest that absence or aberration of the RAIR is a trigger for impaired continence after a low anterior resection [6]. There have been conflicting views in the literature regarding the RAIR after surgical approaches: (1) passive FI with increased recovery of the RAIR after rectal excision [115] and (2) high prevalence of nocturnal FI in patients without a RAIR post-restorative proctocolectomy and ileoanal anastomosis [116]. Zbar et al. [117] separated the RAIR into different sphincter segments in normal controls, constipated patients and patients with FI. They revealed significant linear trends for most parameters at each sphincter level. Recovery time and area under the inhibitory curve differed between the sphincter levels and patient groups, with the most rapid recovery occurring in the distal sphincter of incontinent patients.

4.2.5 Vascular Cushions

Another key player in the maintenance of continence is the anal vascular cushions [118, 119]. The anal vascular cushions, which also include the superior haemorrhoidal plexus, contribute about 15% towards the resting anal tone [30] and provide what is called the 'hermetic seal' that is to fill spaces that cannot be occupied by the sphincteric musculature solely [11]. The vascular cushions have the extraordinary capacity to expand, in order to allow the anal canal to close and preserve continence, when anal pressure decreases. The clinical significance of the vascular cushions becomes apparent, when patients present with FI after haemorrhoidectomy, even with normal sphincter pressure. The mucocutaneous junction, which is the high-pressure zone of the anal canal, creates a barrier that prevents the loss of mucus and faeces. In patients with haemorrhoidal prolapse, this junction is displaced out of the anal margin, causing involuntary faecal or mucus leakage, and thus correction of the prolapse is needed to preserve continence [120].

4.2.6 Stool Consistency

Stool volume and consistency are considered to be also important contributing factors in maintaining continence [8], which directly relates to colonic transit time [121]. The synchronisation of colonic motor activity drives transit, and hence the rate at which colonic contents are transported to the rectum, including the physical and chemical quality of the faeces itself [11]. The colonic transit time is generally rapid when the large stool content is liquid because the left colon does not store fluid well [8, 122, 123]. To ensure normal faecal control one should ask whether the faecal contents are solid, liquid or gas, as some patients may be continent for solid stool, but not necessarily for liquid or gas. Stool can build up in the rectum for a variable time period before the urge to defaecate is experienced. The ability of the rectum to retain stool is known as reservoir continence. The lateral angulations found in the sigmoid colon and the valves of Houston provide a mechanical barrier and retard progression of stool. It is the weight of the stool that tends to accentuate these angles and thus enhance their barrier effect [8]. The high-pressure zone found in the anal canal and the anorectal angle offer a mechanical barrier to defaecation [124] and the rectal curvatures and transverse folds as mentioned above may also contribute to a lesser extent [47].

In patients with constipation slow colonic transit and reduced motility have been found [80]. Degen and Phillips et al. (1996) [122] used scintigraphy and radio-opaque markers in 32 healthy adults, demonstrating the association of hard stools with slower intraluminal movement and loose stool with faster transit. Comparison studies have demonstrated uncoordinated contractile activity in the pelvic colon in patients with constipation compared to those with diarrhoea [125]. This indicates that both reduced colonic motor activity and delayed transit permit more water absorption from intraluminal contents leading to dry stool and reducing volume, subsequently harder stools causing evacuation difficulty [11]. Benninga et al. (1996) [126] found a significant relationship between the presence of a palpable rectal mass and colonic transit time of more than 100 h in children with constipation, who suffered from nocturnal 'overflow' FI. The correlation with stool form and whole gut and colonic transit remains controversial in the literature. In patients with constipation, there has been a correlation with hard stool form and delayed whole gut and colonic transit [123] and this relationship can be absent in healthy subjects [127]. However, there has been a poor correlation with whole gut or colonic transit with stool form and frequency of bowel motions in both adults [123, 127] and children [128]. Thus, true slow transit is generally related to stool infrequency, yet frequent bowel motions do not necessarily imply fast transit and a patient with constipation needs to revisit the toilet frequently [129]. The effect of stool size and consistency on defaecation has shown that evacuation of small hard spheres mimicking pellet-like stool needs more effort than the expulsion of a compressible 50 ml balloon, used as a substitute of soft stool [78].

4.2.7 Rectal Compliance

The normal rectum has the ability to accommodate increases in volume even with a slight change of pressure and it is the adaptation reaction which causes the delay in the passage of the intestinal contents [38]. This reaction prevents the changes in rectal volume that will cause a rise in rectal pressure, which will compromise continence. Electromanometric studies have demonstrated the steep rise in pressure followed by a slow pressure decrease at rest. This change of volume per unit of change of pressure is what is known as rectal compliance [38]. Rectal compliance is prominent at lower volumes of rectal filling expressing active rectal relaxation to accommodate rectal contents [130]. Even with the slight alteration of rectal pressure, the rectum has the ability to adapt, by sensory perception of anorectal activity, which relies on both luminal content and state of contractility or compliance of the rectum [6].

Altered rectal compliance may result in enhanced or diminished rectal tone, either of which may lead to impaired rectal function. It has been suggested that increased compliance results in decreased rectal tone and decreased rectal contractility with possible contribution to delayed evacuation of faeces [131] and decreased rectal compliance might lead to patients having frequent defaecation or FI [132]. The barostat is a physiological tool that has been used to measure rectal compliance by using a catheter attached to an intraluminal balloon, allowing the barostat to deliver fixed-pressure distension to the rectum while measuring alterations in volume, tone and sensory thresholds [6, 133]. Studies have demonstrated the association of altered rectal compliance and patients with faecal urgency, incontinence and constipation [42, 134–136]; there remains inconsistency in the interpretation of these measurements, due to a lack of standardised protocols. The alteration of rectal compliance can be due to various factors including abnormalities of (1) rectal sensation, (2) rectal wall contractility or (3) combination of both [6]. For instance, heightened sensitivity, increased contractility or reduced rectal compliance has been shown in patients with FI, proctitis, pouchitis [42, 135] and scleroderma [137] and as a long-term consequence post-pelvic radiotherapy [138, 139]. In irritable bowel syndrome rectal compliance has been shown to be both lower and unchanged [140]. Conversely, in patients with constipation or megacolon, some may have increased rectal compliance [38, 136] again possibly as the result of reduced rectal sensitivity [141].

4.3 Mechanism of Defaecation

This section focuses on the mechanism of defaecation—key players that have been previously explained in the mechanism of continence will be revisited with their role during human defaecation. The process of defaecation, just as with continence, yet more complex, requires the interplay and coordination of the pelvic musculature which are mediated by the neuromotor and sensorimotor instincts.

The amount of times one defaecates has been used as a guidance to define constipation; however studies have shown that what is considered as 'normal' frequency of defaecation remains variable [125, 142]. The discrepancy of what is considered normal frequency of defaecation is greatly influenced by perceptions from the patient compared to the healthcare professional [143–145]. It is universally accepted that the 'normal' frequency ranges between a maximum of three times per day and minimum of three times per week [146]. The Rome IV Criteria have a clear definition of how to define a patient who is constipated—which is based on not only frequency, but also factors such as straining, stool consistency, feel of incomplete evacuation or obstruction [91]. Studies have explored the variation of frequency with age: motions are rapid at the intial stage of birth; it decreases as the child reaches 3 years old; and by the age of 4 years, the decline rapidly occurs, where the child reaches defaecation frequency similar to an adult [128, 147, 148].

Prior to embarking the physiology of defaecation, it is important to recognise that there are several factors that will influence this mechanism. Studies have shown a connection of higher incidence of constipation in patients with psychological impairment, history of trauma and mental state (e.g. anxiety) [149–154]. Clayden and Agnarsson (1991) [29] discussed the 'withholding' behaviour that is so commonly observed in children [155–161], who are highly motivated to avoid defaecation because of a historical experience of the stool being forced through an incompletely relaxed anorectum by the rectal contractions at maximal rectal loading which leads to rectal pain, bleeding, anal fissure and more worryingly fear of defaecation. Posture has been widely recognised as a significant factor which influences defaecation. Studies have demonstrated that the anorectal angle opens up more easily, with increasing hip flexion, making the evacuation process easier [162]. The squatting posture compared to sitting up straight or lying down [163–165] has been shown to be the superior posture and furthermore minimises disordered defaecation dynamics (i.e. dyssynergia) due to uncoordinated pelvic floor activity. Other factors such as stool consistency have been discussed in previous sections; type of diet and fluid intake, age and gender, cognitive ability, mobility, access of sanitation, cultural and lifestyle factors and hormonal imbalance all influence defaecation [11], but are beyond the scope of this chapter to discuss further.

4.3.1 Defaecation Mechanisms: Step by Step

The mechanism of defaecation begins with the more proximal ileo-colonic motility—that is, the sensation of 'call to stool' and terminate with the opening of the anal sphincters and the expulsion of stools [166]. The physiological mechanism of defaecation is rather complex and best explained in four phases adopted by Palit and Scott (2012) [11]: (1) basal phase; (2) pre-defaecatory phase and urge; (3) expulsive phase and where evacuation occurs; and finally (4) termination of defaecation. Refer to Fig. 4.2 throughout each phase.

4.3.1.1 Phase 1: The Basal Phase

Various reflexes such as the gastrocolic and ileocolic reflexes are involved in the contraction of the colon which leads to stomach, ileum and rectal filing with colonic contents [38].

Firstly, colonic activity is where colonic contents are transported by colonic motility patterns. The transportation begins with the absorption of water from intraluminal contents, the propulsion of the contents at an adequate rate and a place for faeces to temporarily store, until it is physiological and socially acceptable to expel it [11]. A circadian rhythm influences the colonic motor activity where it has been demonstrated that activity increases after awakening—in fact higher in the day than night and after meals [163, 164]. Various diagnostic investigations such as the commonly used radio-opaque markers or the more recent approach wireless telemetric capsule measure the transit (i.e. intraluminal movement) [165–167] and the colonic manometry measures 'contractile activities' (i.e. the sum of which underlies the shift in intraluminal content) which has been used clinically within paediatrics and as a research tool in adults [168–170]. Studies have found in patients with constipation the lack of nocturnal suppression of colonic activity [171], reduced colonic responses to food [172–175] or lack of spatio-temporal organisation of colonic contractile patterns [171].

Shortly after a meal, stool generally moves into the rectum by colonic high-amplitude propagated contractions (HAPC) [176]. The activity of the colon is based on phasic (brief) contractions which can be propagating or non-propagating contractions, or sequences, based on whether or not they propagate along the colon and sustained (tonic) contractions [11]. The role of non-propagated activity in luminal transport is not completely understood [177], but it has been thought to aid mixing of intraluminal contents by local propulsion [178, 179] and retropulsion [180] of the faecal bolus.

The forceful colonic propulsive activity, that is, the HAPCs, is responsible for the shift of large quantity of endoluminal contents in an oro-aboral direction [181, 182]. Thus, the propagated activ-

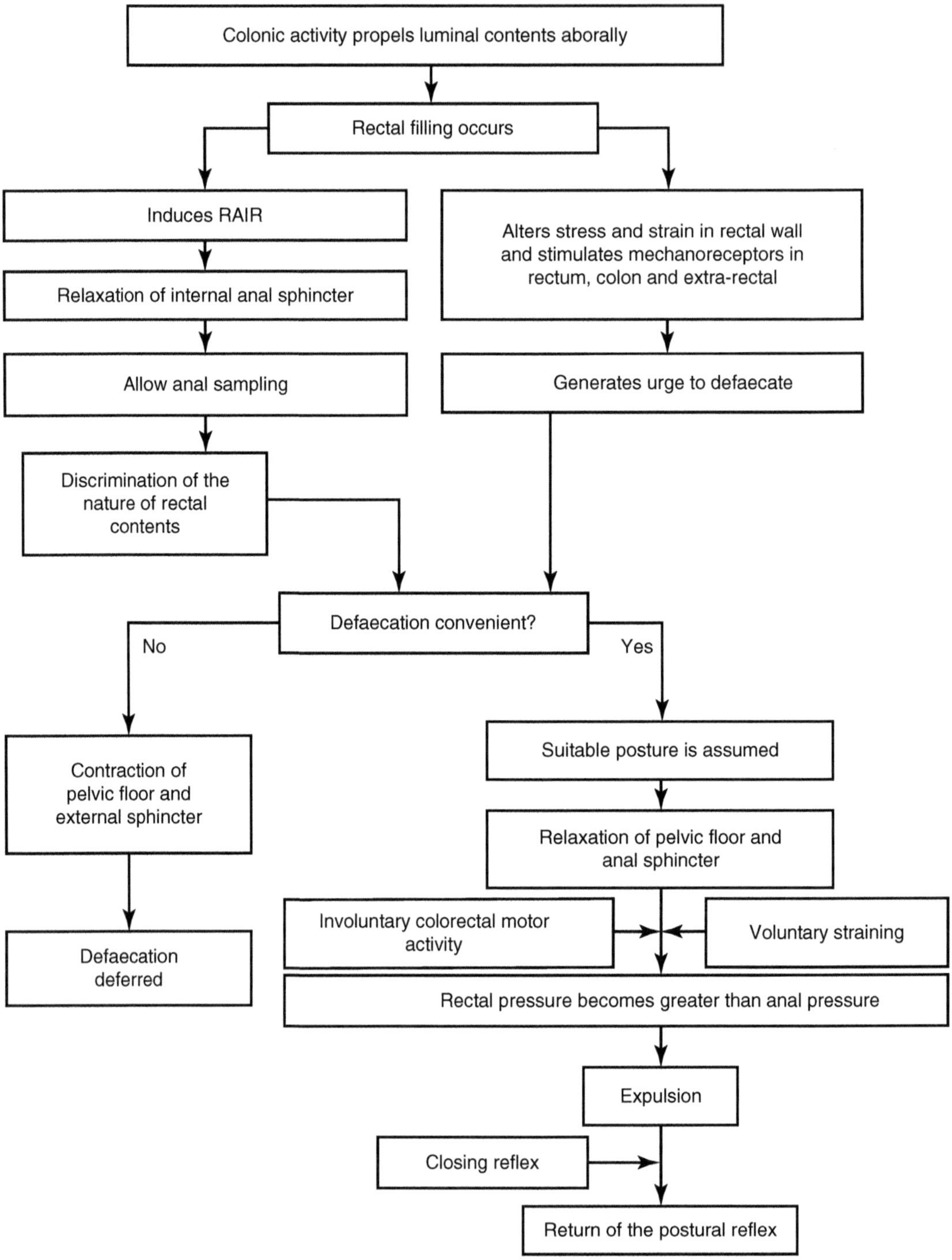

Fig. 4.2 Defaecation events (reprinted with permission) [11]

ity regulated the transfer of contents into the rectum, which is furthermore supported by distension of the sigmoid, which causes contraction simultaneously with relaxation of the recto-sigmoid junction [183]. Studies have explored these contractions and sequences [172] including HAPSs [180, 184–186] in patients with constipation compared to healthy individuals which has been proposed as an important pathophysiological mechanism of delayed colonic transit. HAPSs have been associated with defaecation or passing flatus and their role is to assist in the propulsion of the faecal bolus [163, 185]. Studies have found that the frequency of HAPSs is often reduced in patients with constipation [163, 174] describing motor abnormality in such patients.

Unlike in colonic motor activity, rectal motor activity is unaffected by meal intake [163]. Despite limited understanding of rectal motor activity, they have been observed to propagate in a retrograde direction—that is, they assist to keep the rectum empty by acting as a 'braking mechanism' to the untimely flow of colonic contents [11, 92]. Studies have demonstrated in healthy subjects that during the basal phase, the rectum remains generally empty [187] or it can actually contain a variable quantity of faeces without actual conscious awareness [188].

As mentioned previously the LAM, puborectalis muscles and EAS remain in a state of continuous contraction at rest, which is known as the 'postural reflex' [56]. As the name implies, in human defaecation, when it is socially appropriate, the sitting or squatting position is adopted, the latter position being eased by hip flexion manoeuvre, leading to the optimal straightening of the anorectal angle and allowing more effective propulsion of contents [162]. As we know, the puborectalis muscle maintains the anorectal angle to preserve continence, and increased acuity has been associated to obstructed defaecation [79, 93].

4.3.1.2 Phase 2: The Pre-expulsive Phase—Initiation of Urge to Defaecate

In human defaecation one needs to become aware of the urge or need to defaecate, which is accomplished by several physiological players during this phase. Bampton et al. (2000) [185] demonstrated in the colon that propagated sequences often start as unperceived colonic contractions in the proximal part of the colon, and then travel distally while increasing in amplitude to become a 'full-blown' HAPS, which has been linked to the urge to defaecate.

It is at the rectum that the defaecatory urge starts. As previously mentioned, contraction of the colon, caused by filling of the stomach and abdominal musculature, initiates defaecation [38], which is triggered by sensory response starting with an initial awareness of filling [100]. Once the initial urge is facilitated, constant sensation follows which usually gives the desire to pass flatus and eventually the sustained urge to defaecate arrives, producing that intense urge to defaecate as the maximal tolerable volume/pressure is reached [94, 189, 190].

We appreciate that an intact anal canal sensation is needed for sampling faecal contents (i.e. RAIR), yet it remains uncertain in the literature whether it directly contributes to the generation of a defaecatory urge [11]. Golligher et al. (1951) [191] demonstrated in patients with a colo-anal anastomosis (i.e. anal canal distal to the mucocutaneous junction was preserved) that balloon distension generally elicited a sense of 'wind' or discomfort in the perineum/sacrum, and quite seldom a slight sensation similar to rectal stimulation—a signal that faecal content is on its way.

As mentioned in the first phase, colonic motor activity involves propagated sequences. During the pre-expulsive phase, it has been demonstrated that there is an increase in the frequency and amplitude of propagated sequences [185], which has been found in patients with constipation [192, 193]. Furthermore, studies have found that during the initial phase of antegrade migration, movement of luminal contents distally could possibly motivate distal colonic afferents, which may initiate gradually retrograde propagated sequences in addition to the sensation of urge.

As seen during this phase, segmental and propagated contractions transport colonic content into the rectum, where the sense of rectal filling is pronounced by sensory mechanisms. Rectal filling sensation corresponds with a rise in rectal pressure; however once this is established, the RAIR has been initiated and allows the discrimination of solid from liquid or gaseous luminal contents [194]. Perception of rectal distension can be impaired (i.e. blunted) which is called rectal hyposensitivity and is related to an attenuated 'call to stool' and constipation [151, 195–198], with or without overflow incontinence [94, 110, 151, 199, 200]. Equally, increased perception of distension known as rectal hypersensitivity is linked with a heightened sense of urge, linked to the symptom of faecal urgency, with or without incontinence [201–204].

Conflicting views remain in the literature, regarding the clinical importance of impaired peripheral sensations within paediatrics. Some studies have demonstrated habitual suppression of the defaecatory urge, which diminished the call to stool, leading to faecal impaction and secondary dilation or megarectum [205–207]. However, other recent studies have argued that there is actually no differences in sensory function in children with functional constipation, compared to healthy subjects [134], but other pathophysiological mechanisms were found in these patients to be involved such as greater rectal compliance. Others have postulated other mechanisms involved in children with idiopathic constipation or megarectum including neuromuscular dysfunctions [12, 208, 209] and psychological [210], behavioural and neurophysiological factors [205].

As the rectum distends, stretch receptors in the rectal walls are activated, and as the stool descends even more receptors become involved and an increase of sensation occurs. As discussed, rectal sensation expresses the feeling of rectal filling along with anal reflexes. In vivo studies have demonstrated that mechanoreceptors act slowly to respond to tension and rapid distension in the colon [211, 212]. Others have postulated that in vivo, the incoming faecal bolus distorts the rectal wall, altering stress and strain, and thus activating mechanoreceptors that then prompt reflex rectal contractions [213, 214]. These tension mechanoreceptors respond to both rectal distension and muscle contraction which relies on rectal filling sensation and period of raised rectal pressure during rectal distension [194, 215]. Immunohistochemical studies have also recognised mucosal afferents that are both mechano- and chemosensitive and have been associated with increased numbers in patients with rectal hypersensitivity [212, 216].

Is volume, pressure or weight of rectal contents the main trigger for rectal sensation? Broens et al. (1994) [189] used a rectal balloon with 60ml of air, water and mercury to test this and found that sensation levels were independent of both weight and volume of the rectal contents. They postulated that rectal sensation is actually sensitive to intrarectal pressure changes which trigger tension-activated stretch receptors. Yet, others have suggested that the direct stimulus is rectal wall deformation rather than intrarectal pressure, since mechanoreceptors are influenced by strain and force that cause changes in rectal wall morphology [217–219].

The perception of rectal sensation has been found to be mainly controlled by sacral outflow with less involvement from the thoracolumbar outflow [11]. This has been shown where rectal sensation has been preserved after bilateral pudendal nerve block [22, 220], yet low spinal anaesthesia (i.e. L1–S1) eliminates rectal sensation, and thus is perceived only as a vague abdominal discomfort at higher levels of rectal filling, which has been shown in high spinal anaesthesia (T6–T12) [32]. Cerebral evoked potentials have been measured in response to rectal balloon distension. Loening Baucke et al. (1995) [221] demonstrated that children with chronic constipation and encopresis have profoundly prolonged latencies indicative of a defect in the afferent pathway from the rectum.

The capability of the rectum to accommodate increasing volume with limited alterations to pressure [6] is known as compliance and has been previously mentioned in this chapter. Rectal compliance is most distinct at lower volumes of rectal filling, which demonstrates adaptive relaxation [141] and a temporary storage of the rectal contents [130]. Therefore, sensory perception of anorectal activity depends on both luminal content and state of contractility, or compliance, of the rectum. Other properties such as viscoelasticity of the rectal wall have been suggested to influence rectal sensation [11, 222]. Patients with rectal hyposensitivity could be due to (a) abnormal rectal wall properties where afferent nerve function may be intact or (b) impaired afferent function which can occur at any level of the pathway from receptor to higher centres of the central nervous system [11, 223].

When we get the urge to defaecate, the EAS and the pelvic floor muscles can be further voluntarily contracted [56, 191], increasing the acuity of the anorectal angle, elevating the pelvic floor and lengthening the high-pressure zone of the anal canal [224]. Broens et al. (2002) [194] explored the relationships of the anal canal relaxation allied to rectal filling sensation. It was demonstrated that the filling volume elicited a constant sensation, and the upper anal canal diameter was 3.2 cm, which increased to 4 and 4.4 cm at urge and maximum tolerable volumes, respectively.

4.3.1.3 Phase 3: The Expulsive Phase—Evacuation

The sampling reflex (i.e. RAIR) and the defaecatory urge simultaneously work together during the expulsive phase and a conscious decision is made to defaecate. Evacuation is influenced by voluntary straining and appropriate posture, which is based on the elevation in intrarectal pressure and relaxation of the pelvic floor musculature and anal canal. Defaecation may be voluntarily suppressed, based on the physical nature and volume of the stool [11].

Most of the colon and rectum are empty during defaecation. This was shown in a scintigraphic study [225], where defaecation in 11 healthy volunteers demonstrated that the mean percentage of segmental evacuation was right colon 20%, left colon 32% and rectum 66%. In order for faecal contents to be evacuated, an increase in intrarectal pressure is required [84, 222] in conjunction with relaxation of the anal canal resulting in decreased anal pressure. Additionally, during defaecation straining occurs, which increases intra-pelvic and intrarectal pressure.

Some studies have shown that evacuation is not accompanied with rectal contraction [226] and others show that there is no noticeable rise in intrarectal pressure in relation to intra-pelvic pressure during evacuation [227]. It has been postulated that evacuation is in fact effected by variable factors including voluntary straining and cooperative colorectal contractions [84] which are further influenced by stool volume/consistency and behavioural and cultural habits (i.e. timing of defaecation in relation to the onset of defaecatory urge) [11].

A reflex inhibition of the pelvic floor has been suggested as part of this phase, where there is a rise in abdominal pressure (stretch stimulus), which is initially excitatory to the pelvic floor, but later inhibitory when prolonged beyond a critical window [56]. A 'gating mechanism' has been proposed recently, which allows or prevents stimuli from various sources (e.g. increased intra-abdominal pressure, pelvic organ distension) to excite or inhibit the motor neurons [228]. Evacuation with ease is mediated by pelvic floor relaxation, in addition to high intra-abdominal pressure which causes it to descend [79]; otherwise defaecation dynamics is compromised (e.g. pelvic floor dyssynergia, or dyssynergic defaecation) [12, 79, 83, 224, 229, 230]. The anorectal angle straightens due to relaxation of the puborectalis part of the pelvic floor, which is assisted by posture, involving a degree of hip flexion as previously discussed [11].

The anal canal and sophisticated sphincters also relax during this phase and if this does not occur, pelvic floor dyssynergia has been found in both adult and paediatric cohorts [12, 83, 124, 229]. Studies have shown that failure coordinates the increased intra-abdominal pressure with adequate pelvic floor relaxation, and it has been shown in infants who are constipated which may be triggered by inappropriate contraction of the EAS during defaecation which can be related to behavioural or faecal retention [102, 157].

Regarding the IAS during this phase, relaxation occurs involuntarily in response to rectal distension and relaxation is relative to the intrarectal pressure [15, 28]. This is followed by straining, which is performed contracting the abdominal muscles and diaphragm against a closed glottis (i.e. Valsalva manoeuvre), which is associated with relaxation of the EAS. Petros et al. (2008) [230] suggested that simultaneous contraction of the levator plate (that inserts into the posterior aspect of the rectum) and the longitudinal muscles of the anus also occurs during evacuation. This can be explained by the resultant force vector which is directed posteriorly and downwards allowing the anorectal angle to open. This is further supported by the contraction of the pubococcygeus muscle that is able to strap the perineal body, which allows the anterior wall of the anal canal to tense, permitting only the posterior wall to move backwards [230]. Contraction of the longitudinal muscles of the anus [231] and shortening of the anal canal [177] and the faecal bolus itself [232] allow flattening of the vascular cushions. These simultaneous actions allow the decrease of the anal canal pressure (i.e. lower than the intrarectal pressure) ensuring a pressure gradient from the rectum to the outside [11]. Expulsion then occurs and persists due to the high intrarectal pressure, improved by straining.

4.3.1.4 Final Phase: Termination of Defaecation

This phase is initiated with a degree of voluntary control (feeling of complete evacuation), followed by the synchronised manoeuvres to increase in-pelvic pressure and the involuntary contraction of both the EAS and pelvic floor muscles, which lead to the closure of the anal canal [11]. The 'closing reflex' is an important part of this final phase. Where there is the action of something pulling to the anus, that this reflex responds, which allows the EAS to have temporary increase in activity that assists in the closure of the anal canal [6, 56, 177] and mostly important to complete this process, provide the IAS, time to recover its tone [233] (Nyam et al.). It has been suggested that this reflex is influenced cortically, as it has been demonstrated to be impaired in patients with spinal injury [56]. To complete the process, straining stops followed by a decrease in intra-abdominal pressure and the postural reflex is reactivated to allow the contraction of the puborectalis, to return the anorectal angle back to its basal state [11, 56]. Relaxation of the longitudinal muscle simultaneously lengthens the anal canal and anal cushions passively distend, finally leading to the complete closure of the anal canal [11].

4.4 Conclusion

The notion of having a bowel motion is far from simple. In fact it involves a masterly performance that is complex. This chapter has explored the physiological processes involved, including both conscious and subconscious interaction for both continence and defaecation. The synergy of actions from the musculature, brain, spinal cord, peripheral nerves and anatomy and associated factors such as culture, age, environment, psychology, diet and behaviour are all equally significant in the preservation of continence. There is no doubt that our knowledge remains limited, especially within paediatrics; however, further research will enhance our physiological understanding and how to improve treatment modalities.

References

1. Butler Tjaden NE, Trainor PA. The developmental etiology and pathogenesis of Hirschsprung disease. Transl Res. 2013;162:1–15.
2. Versalii Eruxellensis Andreae, de Humani corporis fabrica. De Recti intestine musculis. 1st ed., pp. 228; 1543.
3. Lamb MN, Kaiser AM. Anatomy, physiology, and measurement of physiologic function for colorectal surgery. Sci Am Surg. 2015:1–20.
4. Joseph TN. Clinical gastroenterology, vol. 121. Australia: Elsevier; 2006.
5. Debongnie JC, Phillips SF. Capacity of the human colon to absorb fluid. Gastroenterology. 1978;74:698–703.
6. Bajwa A, Emmanuel A. The physiology of continence and evacuation. Best Prac Res Clin Gastrol. 2009;23:477–85.
7. Ramussen OO, Christiansen J. Physiology and pathophysiology of anal function. Scand J Gastroenterol Suppl. 1996;216:169–74.
8. Gordon PH. Surgical anatomy: Principles and practice of surgery for the colon and rectum and anus. New York: Informa Healthcare; 2007.
9. Rao SS. Pathophysiology of adult fecal incontinence. Gastroenterology. 2004;126:S14–22.
10. Hobday DI, Aziz Q, Thacker N, Hollander I, Jackson A, Thompson DG. A study of the cortical processing of ano-rectal sensation using functional MRI. Brain. 2001;124:361–8.
11. Palit S, Lunniss PJ, Scott SM. The physiology of human defecation. Dig Dis Sci. 2012;57:1445–64.
12. Lunniss P, Gladman M, Benninga M, Rao S. Pathophysiology of evacuation disorders. Neurogastroenterol Motil. 2009;21:31–40.
13. Raizada V, Ravinder M. Pelvic floor anatomy and applied physiology. Gastroenterol Clin North Am. 2008;37(3):493.
14. Standring S. Gray's anatomy: the anatomical basis of clinical practice. Churchill Livingstone London. 2004.

15. Hancock BD. Measurement of anal pressure and motility. Gut. 1976;17:645–51.
16. Pedersen IK, Christiansen J. A study of the physiological variation in anal manometry. Br J Surg. 1989;76:69–70.
17. Gonella J, Bouvier M, Blanquet F. Extrinsic nervous control of motility of small and large intestines and related sphincters. Physiol Rev. 1987;67:902–61.
18. Wankling WJ, Brown BH, Collins CD, Duthie HL. Basal electrical activity in the anal canal in man. Gut. 1968;9:457–60.
19. Garavoglia M, Borghi F, Levei AC. Arrangement of the anal striated musculature. Dis Colon Rectum. 1993;36(1):10–5.
20. Gutierreze JG, Oliai A, Chey WY. Manometric profile of the internal sphincter in man. Gastroenterology. 1975;68:907.
21. Enck P, Eggers E, Koletzko S, Erckenbrecht JF. Spontaneous variation of anal resting pressure in healthy humans. Am J Physiol. 1991;261: G823–6.
22. Frenckner B, Euler CV. Influence of pudendal block on the function of the anal sphincters. Gut. 1975;16:482–9.
23. Ustach TJ, Tobon F, Hambrecht T, Bass DD, Schuster MM. Electrophysiological aspects of human sphincter function. J Clin Invest. 1970;49:41–8.
24. Orkin BA, Hanson RB, Kelly KA, Phillips SF, Dent J. Human anal motility while fasting, after feeding and during sleep. Gastroenterology. 1991;100:1016–23.
25. Read NW, Haynes WG, Bartolo DC, et al. Use of anorectal manometry during rectal infusion of saline to investigate function in incontinent patients. Gastroenterology. 1983;85:105–13.
26. Miller R, Lewis GT, Bartolog DC, Cervero F, Mortensen NJ. Sensory discrimination and dynamic activity in the anorectum evidence of a new ambulatory technique. Br J Surg. 1988;75:44–7.
27. Sangwan YP, Solla JA. Internal anal sphincter: advances and insights. Dis Colon Rectum. 1998;41:1297–311.
28. Frenckner B. Function of the anal sphincters in spinal man. Gut. 1975;16:638–44.
29. Clayden G, Agnarsson U. Constipation in childhood. Oxford Medical Publications; 1991.
30. Lestar B, Penninckx F, Kerremans R. The composition of anal basal pressure. An in vivo and in vitro study in man. Int J Colorectal Dis. 1989;4:118–22.
31. Waldron DJ, Kumar D, Hallan RI, Williams NS. Prolonged ambulant assessment of anorectal function in patients with prolapse haemorrhoids. Dis Colon Rectum. 1989;32:968–74.
32. Frenckner B, Ihre T. Influence of autonomic nerves on the internal anal sphincter in man. Gut. 1976;17:306–12.
33. Penninckx F, Lestar B, Kerremans R. The internal anal sphincter: mechanisms of control and its role in maintaining anal continence. Baillieres Clin Gastroenterol. 1992;6:193–214.
34. Horgan PG, O'Connell PR, Shinkwin CA, Kirwan WO. Effect of anterior resection on anal sphincter function. Br J Surg. 1989;76:783–6.
35. Lubowski DZ, Nicholls RJ, Swash M, et al. Neural control of internal anal sphincter function. Br J Surg. 1987;74:668–70.
36. Bharucha AE, Fletcher JG. Recent advances in assessing anorectal structure and functions. Gastroenterology. 2007;133:1069–74.
37. Dyck PJ, Thomas PK. Autonomic and somatic systems to the anorectum and pelvic floor. 4th ed. Philadelphia PA: Elsvier Saunders; 2005.
38. Holschneider AM, Fritsch H, Holschneider P. Chapter 7: anatomy and function of the normal rectum and anus. In: Holschneider AM, Hutson JM, editors. Anorectal malformations in children. Springer; 2006.
39. Rattan S. The internal anal sphincter: regulation of smooth muscle tone and relaxation. Neurogastroenterol Motil. 2005;17:50–9.
40. Goyal RK, Hirano I. The enteric nervous system. N Engl J Med. 1996;334:1106–15.
41. Engel AF, Kamm MA, Bartram CI, et al. Relationship of symptoms in faecal incontinence to specific sphincter abnormalities. Int J Colorectal Dis. 1995;10:152–5.
42. Salvioli B, Bharucha AE, Rath-Harvey D, et al. Rectal compliance, capacity, and rectoanal sensation in fecal incontinence. Am J Gastroenterol. 2001;96:2158–68.
43. Read NW, Bartolo DC, Read MG. Differences in anal function in patients with incontinence to solids and in patients with incontinence to liquids. Br J Surg. 1984;71:39–42.
44. Walker WA, Rothenberger DA, Goldberg SM. Morbidity of internal sphincterotomy for anal fissure and stenosis. Dis Colon Rectum. 1985;28:832–5.
45. Santorini J. Septemdecim tabulae; 1715
46. Bogduk N. Issues in anatomy: the external anal sphincter revisited. Aust N Z J Surg. 1996;66:626–9.
47. Bharucha AE. Pelvic floor: anatomy and function. Neurogastroenterol Motil. 2006;18:507–19.
48. Stoker J, Halligan S, Bartram CI. Pelvic floor imaging. Radiology. 2001;218:621–41.
49. Ayoub SF. Anatomy of the external anal sphincter in man. Acta Anat (Basel). 1979;105:25–36.
50. Shafik A. A new concept of the anatomy of the anal sphincter mechanism and the physiology of defecation. The external anal sphincter: a triple-loop system. Invest Urol. 1975;12:412–9.
51. Peschers UM, DeLancey JO, Fritsch H, Quint LE, Prince MR. Cross-sectional imaging anatomy of the anal sphincters. Obstet Gynecol. 1997;90(5): 839–44.
52. Lunniss PJ, Scott SM. Imaging of the anal sphincter. In: Sultan AH, Thakar R, Fenner DE, editors. Pathophysiology of anal incontinence. London: Springer; 2007.
53. Floyd WF, Walls EW. Electromyography of the sphincter ani externus in man. J Physiol. 1953;121:49P–50P.

54. Jorge JMN, Wexner SD. A practical guide to basic anorectal physiology. Constemp Surg. 1993; 43:214–24.
55. Jorge JMN, Habr-Gama. The value of sphincter asymmetry index in anal incontinence. Int J Color Dis. 2000;15(5–6):303–10.
56. Porter NH. A physiological study of the pelvic floor in rectal prolapse. Ann R Coll Surg Engl. 1962:31–379.
57. Taverner D, Smiddy FG. An electromyographic study of the normal function of the external anal sphincter and pelvic diaphragm. Dis Colon Rectum. 1959;2:153–60.
58. Kumar D, Waldron D, Williams NS, Browning C, Hutton MR, Wingate DL. Prolonged anorectal manometry and external anal sphincter electromyography in ambulant human subjects. Dig Dis Sci. 1990;35:641–8.
59. Duthie HL, Watts J. Contribution of the external anal sphincter to the pressure zone in the anal canal. Gut. 1965;6:64–8.
60. Parks AG, Porter NH, Melzak J. Experimental study of the reflex mechanisms controlling the muscles of the pelvic floor. Dis Colon Rectum. 1962;5:407–14.
61. Sultan AH, Kamm MA, Hudson CN, et al. Anal-sphincter disruption during vaginal delivery. N Engl J Med. 1993;329:1905–11.
62. Rintala R, Lindahl H, Marttinen E, Sariola H. Constipation is a major functional complication after internal sphincter-saving posterior sagittal anorectoplasty for high and intermediate anorectal malformations. J Pediatr Surg. 1993;28:1054–8.
63. Rintala RJ, Lindahl H. Is normal bowel function possible after repair of intermediate and high anorectal malformations? J Pediatr Surg. 1995;30:491–4.
64. Ong NT, Beasley SW. Long-term continence in patients with high and intermediate anorectal anomalies treated by sacroperineal (Stephens) rectoplasty. J Pediatr Surg. 1991;26:44–8.
65. Athanasakos EP, Ward HC, Williams NS, Scott SM. The importance of extrasphincteric mechanisms in the pathophysiology of faecal incontinence in adults with history of anorectal anomaly. Br J Surg. 2008;95:1394–400.
66. Gaston EA. Physiological basis for preservation of fecal continence after resection of rectum. JAMA. 1951;146:1486–9.
67. Dujovny N, Quiros RM, Saclarides TJ. Anorectal anatomy and embryology. Surg Oncol Clin N Am. 2004;13:277–93.
68. Kaiser AM, Ortega AE. Anorectal anatomy. Surg Clin North Am. 2002;82:1125–38.
69. Parks AG. Anorectal incontinence. Proc Roy Soc Med. 1975;68:61–90.
70. Nielsen MB, Hauge C, Rasmussen O, Pedersen JF, Christiansen J. Anal endosonographic findings in the follow-up of primary sutured sphincteric ruptures. Br J Surg. 1992;79:104–6.
71. Oliveira LCC, Salum MR, Saad RC. Chapter 2 Physiology of continence and defecation. In: Anorectal physiology: a clinical and surgical perspective. 1st ed. Springer; 2020.
72. Singh K, Jakab M, Reid WM, Berger LA, Hoyte L. Three-dimensional magnetic resonance imaging assessment of levator ani morphologic features in different grades of prolapse. Am J Obstet Gynecol. 2003;188:910–5.
73. Bharucha AE, Fletcher JG, Seide B, Riederer SJ, Zinsmeister AR. Phenotypic variation in functional disorders of defecation. Gastroenterology. 2005;128:1199–210.
74. Bharucha AE, Fletcher JG, Harper CM, et al. Relationship between symptoms and disordered continence mechanisms in women with idiopathic faecal incontinence. Gut. 2005;54:546–55.
75. Fletcher JG, Busse RF, Riederer SJ, et al. Magnetic resonance imaging of anatomic and dynamic defects of the pelvic floor in defecatory disorders. Am J Gastroenterol. 2003;98:399–411.
76. Mahieu P, Pringot J, Bodart P. Defecography: I. Description of a new procedure and results in normal patients. Gastrointest Radiol. 1984;9:247–51.
77. Bartolo DC, Read NW, Jarratt JA, Read MG, Donnelly TC, Johnson AG. Differences in anal sphincter function and clinical presentation in patients with pelvic floor descent. Gastroenterology. 1983;85:68–75.
78. Bannister JJ, Davison P, Timms JM, Gibbons C, Read NW. Effect of stool size and consistency on defecation. Gut. 1987;28:1246–50.
79. Bartolo DC, Roe AM, Virjee J, Mortensen NJ. Evacuation proctography in obstructed defaecation and rectal intussusception. Br J Surg. 1985;72:S111–6.
80. Bharucha AE. Lower gastrointestinal functions. Neurogastroenterol Motil. 2008;20:103–13.
81. Parks AG. Royal Society of Medicine, Section of Proctology; Meeting 27 November 1974. President's Address. Anorectal incontinence. Proc R Soc Med. 1975;68:681–90.
82. Percy JP, Neill ME, Swash M, Parks AG. Electrophysiological study of motor nerve supply of pelvic floor. Lancet. 1981;1:16–7.
83. Bharucha AE, Wald A, Enck P, Rao S. Functional anorectal disorders. Gastroenterology. 2006;130:1510–8.
84. Morgan CN. The surgical anatomy of the ischiorectal space. Proc R Soc Med. 1949;42:189–200.
85. Varma KK, Stephens D. Neuromuscular reflexes of rectal continence. Aust N Z J Surg. 1972;41:263–72.
86. Sun WM, Read NW, Donnelly TC. Impaired internal anal sphincter in a subgroup of patients with idiopathic fecal incontinence. Gastroenterology. 1989;97:130–5.
87. Fernandez-Fraga X, Azpiroz F, Malagelada JR. Significance of pelvic floor muscles in anal incontinence. Gastroenterology. 2002;123:1441–50.
88. Broens PMA, Jonker JE, Trzpis. The puborectalis continence reflex: a new regulatory mechanism

controlling fecal continence. Int J Colorectal Dis. 2018;33:627–33.
89. Patton V, Kumar S, Parkin K, Karantais E, Dinning P. The relationship between residual sphincter damage after primary repair, faecal incontinence, and anal sphincter function in primiparous women with an obstetric anal sphincter injury. Neurourol Urodyn. 2019;38(1):193–9.
90. Dietz HP. Quantification of major morphological abnormalities of the levator ani. Ultrasound Obstet Gynecol. 2007;29:329–34.
91. Rao SS, Bharucha AE, Chiarioni G, et al. Functional anorectal disorders. Gastroenterology. 2016;150(6):1430–42.
92. Rao S, Welcher K. Periodic rectal motor activity: the intrinsic colonic gatekeeper? Am J Gastroenterol. 1996;91:890–7.
93. Bannister JJ, Timms JM, Barfield LJ, Donnelly TC, Read NW. Physiological studies in young women with chronic constipation. Int J Colorectal Dis. 1986;1:175–82.
94. Sun WM, Read NW, Miner PB. Relation between rectal sensation and anal function in normal subjects and patients with faecal incontinence. Gut. 1990;31:1056–61.
95. Puri P. Chapter 15: Hirschsprung Disease and Anorectal Manometry. In: Hirschsprung's disease and allied disorders. 4th ed. Springer; 2019. p. 233.
96. KEIGHLEY. Anatomy and physiology: Surgery of the anus, rectum and colon. London: WB Saunder Company; 1993.
97. Duthie HL, Gairns FW. Sensory nerve endings and sensation in the anal region of man. Br J Surg. 1960;47:585–95.
98. Lane RH, Parks AG. Function of the anal sphincters following colo-anal anastomosis. Br J Surg. 1977;64:596–9.
99. Sagar PM, Holdsworth PJ, Johnston D. Correlation between laboratory findings and clinical outcome after restorative proctocolectomy: serial studies in 20 patients with end-to-end pouch-anal anastomosis. Br J Surg. 1991;78:67–70.
100. Meunier P, Mollard P. Control of the internal anal sphincter (manometric study with human subjects). Pflugers Arch. 1977;370:233–9.
101. Beuret-Blanquart F, Weber J, Gouverneur JP, Demangeon S, Denis P. Colonic transit time and anorectal manometric anomalies in 19 patients with complete transection of the spinal cord. J Auton Nerv Syst. 1990;30:199–207.
102. Meunier P, Marechal JM, Mollard P. Accuracy of the manometric diagnosis of Hirschsprung's disease. J Pediatr Surg. 1978;13:411–5.
103. Heij HA, de Vries X, Bremer I. Long-term anorectal function after Duhamel operation for Hirschsprung's disease. J Pediatr Surg. 1995;30:430–2.
104. Nagasaki A, Ikeda K, Suita S. Postoperative sequential anorectal manometric study of children with Hirschsprung's disease. J Pediatr Surg. 1980;15:615–9.
105. Suzuki H, Tsukamoto Y, Amano S, Sakarai A. Motility of the anorectum after 'rectoplasty with posterior triangular colonic flap' in Hirschsprung's disease. Jpn J Surg. 1984;14:335–8.
106. Morikawaw Y, Matsufugi H, Hirobe S, Yokoyama J, Katsumata K. Motility of the anorectal after the Soave-Denda operation. Prog Pediatr Surg. 1989;4:67–76.
107. Van Leeuwen K, Geiger JD, Barnett JL, Coran AG. Teitelbaum DH Stooling and manometric findings after primary pull-throughs in Hirschsprung's disease: perineal versus abdominal approaches. J Pediatr Surg. 2002;37:1321–5.
108. Callaghan RP, Nixon HH. Megarectum: physiological observations. Arch Dis Child. 1964;39:153–7.
109. Nixon HH, Callaghan RP. Ano-rectal anomalies: physiological considerations. Arch Dis Child. 1964;39:158–60.
110. Lubowski DZ, Nicholls RJ. Faecal incontinence associated with reduced pelvic sensation. Br J Surg. 1988;75:1086–8.
111. Rasmussen OO, Petersen IK, Christiansen J. Anorectal function following low anterior resection. Colorectal Dis. 2003;5:258–61.
112. Matzel KE, Stadelmaier U, Muehldorfer S, et al. Continence after colorectal reconstruction following resection: impact of level of anastomosis. Int J Colorectal Dis. 1997;12:82–7.
113. Lee SJ, Park YS. Serial evaluation of anorectal function following low anterior resection of the rectum. Int J Colorectal Di. 1998;13:241–6.
114. Lewis WG, Holdsworth PJ, Stephenson BM, et al. Role of the rectum in the physiological and clinical results of coloanal and colorectal anastomosis after anterior resection for rectal carcinoma. Br J Surg. 1992;79:1082–6.
115. van Duijvendijk P, Slors F, Taat CW, et al. A prospective evaluation of anorectal function after total mesorectal excision in patients with a rectal carcinoma. Surgery. 2003;133:56–65.
116. Saigusa N, Belin BM, Choi HJ, et al. Recovery of the rectoanal inhibitory reflex after restorative proctocolectomy: does it correlate with nocturnal continence? Dis Colon Rectum. 2003;46:168–72.
117. Zbar A, Aslam M, Gold DM, Gatzen C, Gosling A, Kmiot WA. Parameters of the rectoanal inhibitory reflex in patients with idiopathic fecal incontinence and chronic constipation. Dis Colon Rectum. 1998;41:200–8.
118. Gibbons CP, Bannister JJ, Trowbridge EA, Read NW. An analysis of anal sphincter pressure and anal compliance in normal subjects. Int J Color Dis. 1986;1:231–7.
119. Gibbons CP, Trowbridge EA, Bannister JJ, Read NW. Role of anal cushions in maintaining continence. Lancet. 1989;1:886–8.
120. Murie JA, Sim AJ, Mackenzie I. The importance of pain, pruritus and soiling as symptoms of heamorrhoids and their response to haemorrhoidectomy or rubber band ligation. Br J Surg. 1981;68:247–9.

121. Degen LP, Phillips SF. How well does stool form reflect colonic transit? Gut. 1996;39:109–13.
122. Davies GJ, Crowder M, Reid B, Dickerson JW. Bowel function measurements of individuals with different eating patterns. Gut. 1986;27:164–9.
123. O'Donnell LJ, Virjee J, Heaton KW. Detection of pseudodiarrhoea by simple clinical assessment of intestinal transit rate. BMJ. 1990;300:439–40.
124. Rasmussen OO. Anorectal function. Dis Colon Rectum. 1994;37:386–403.
125. Connell AM. The motility of the pelvic colon. II. Paradoxical motility in diarrhoea and constipation. Gut. 1962;3:342–8.
126. Benninga MA, Buller HA, Tytgat GN, Akkermans LM, Bossuyt PM, Taminiau JA. Colonic transit time in constipated children: does pediatric slow-transit constipation exist? J Pediatr Gastroenterol Nutr. 1996;23:241–51.
127. Saad RJ, Rao SS, Koch KL, et al. Do stool form and frequency correlate with whole-gut and colonic transit? Results from a multicenter study in constipated individuals and healthy controls. Am J Gastroenterol. 2010;105:403–11.
128. Corazziari E, Cucchiara S, Staiano A, et al. Gastrointestinal transit time, frequency of defecation, and anorectal manometry in healthy and constipated children. J Pediatr. 1985;106:379–82.
129. Dinning PG, Hunt L, Lubowski DZ, Kalantar JS, Cook IJ, Jones MP. The impact of laxative use upon symptoms in patients with proven slow transit constipation. BMC Gastroenterol. 2011; 11:121.
130. Fox M, Thumshirn M, Fried M, et al. Barostat measurement of rectal compliance and capacity. Dis Colon Rectum. 2006;49:360–70.
131. Mertz H, Naliboff B, Mayer E. Physiology of refractory chronic constipation. Am J Gastroenterol. 1999;94:609–15.
132. Crowell MD. Pathogenesis of slow transit and pelvic floor dysfunction: from bench to bedside. Rev Gastroenterol Disord. 2004;4(Suppl 2):S17–27.
133. van der Berg MM, Di Lorezo C, van Ginkel MJM, Benninga MC. Barostat testing in children with functional gastrointestinal disorders. Curr Gastroenterol Reports. 2006;8:222–7.
134. van den Berg MM, Voskuijl WP, Boeckxstaens GE, Benninga MA. Rectal compliance and rectal sensation in constipated adolescents, recovered adolescents and healthy volunteers. Gut. 2008;57:599–603.
135. Felt-Bersma RJ, van Baren R, Koorevaar M, et al. Unsuspected sphincter defects shown by anal endosonography after anorectal surgery. A prospective study. Dis Colon Rectum. 1995;38:249–53.
136. Gosselink MJ, Hop WC, Schouten WR. Rectal compliance in females with obstructed defecation. Dis Colon Rectum. 2001;44:971–7.
137. Leighton JA, Valdovinos MA, Pemberton JH, et al. Anorectal dysfunction and rectal prolapse in progressive systemic sclerosis. Dis Colon Rectum. 1993;36:182–5.
138. Berndtsson I, Lennernas B, Hulten L. Anorectal function after modern conformal radiation therapy for prostate cancer: a pilot study. Tech Coloproctol. 2002;6:101–4.
139. Yeoh EE, Holloway RH, Fraser RJ, et al. Anorectal dysfunction increases with time following radiation therapy for carcinoma of the prostate. Am J Gastroenterol. 2004;99:361–9.
140. Park JH, Baek YH, Park DI, et al. Analysis of rectal dynamic and static compliances in patients with irritable bowel syndrome. Int J Colorectal Dis. 2008;23:659–64.
141. Gladman MA, Dvorkin LS, Lunniss PJ, Williams NS, Scott SM. Rectal hyposensitivity: a disorder of the rectal wall or the afferent pathway? An assessment using the barostat. Am J Gastroenterol. 2005;100:106–14.
142. Heaton KW, Radvan J, Cripps H, Mountford RA, Braddon FE, Hughes AO. Defecation frequency and timing, and stool form in the general population: a prospective study. Gut. 1992;33:818–24.
143. Herz MJ, Kahan E, Zalevski S, Aframian R, Kuznitz D, Reichman S. Constipation: a different entity for patients and doctors. Fam Pract. 1996;13:156–9.
144. Sandler RS, Jordan MC, Shelton BJ. Demographic and dietary determinants of constipation in the US population. Am J Public Health. 1990;80:185–9.
145. Sandler RS, Drossman DA. Bowel habits in young adults not seeking health care. Dig Dis Sci. 1987;32:841–5.
146. Schaefer DC, Cheskin LJ. Constipation in the elderly. Am Fam Physician. 1998;58:907–14.
147. Fontana M, Bianchi C, Cataldo F, et al. Bowel frequency in healthy children. Acta Paediatr Scand. 1989;78:682–4.
148. Weaver LT. Bowel habit from birth to old age. J Pediatr Gastroenterol Nutr. 1988;7:637–40.
149. Drossman DA, Talley NJ, Leserman J, Olden KW, Barreiro MA. Sexual and physical abuse and gastrointestinal illness. Review and recommendations. Ann Intern Med. 1995;123:782–94.
150. Drossman DA. Abuse, trauma, and GI illness: is there a link? Am J Gastroenterol. 2011;106:14–25.
151. Wald A, Hinds JP, Caruana BJ. Psychological and physiological characteristics of patients with severe idiopathic constipation. Gastroenterology. 1989;97:932–7.
152. Nehra V, Bruce BK, Rath-Harvey DM, Pemberton JH, Camilleri M. Psychological disorders in patients with evacuation disorders and constipation in a tertiary practice. Am J Gastroenterol. 2000;95:1755–8.
153. Dykes S, Smilgin-Humphreys S, Bass C. Chronic idiopathic constipation: a psychological enquiry. Eur J Gastroenterol Hepatol. 2001;13:39–44.
154. Leroi AM, Bernier C, Watier A, et al. Prevalence of sexual abuse among patients with functional disorders of the lower gastrointestinal tract. Int J Colorectal Dis. 1995;10:200–6.
155. Borowitz SM, Cox DJ, Tam A, Ritterband LM, Sutphen JL, Penberthy JK. Precipitants of constipa-

tion during early childhood. J Am Board Fam Pract. 2003;16:213–8.
156. Loening-Baucke V. Constipation in early childhood: patient characteristics, treatment, and long-term follow up. Gut. 1993;34:1400–4.
157. Loening-Baucke VA, Cruikshank BM. Abnormal defecation dynamics in chronically constipated children with encopresis. J Pediatr. 1986;108:562–6.
158. Partin JC, Hamill SK, Fischel JE, Partin JS. Painful defecation and fecal soiling in children. Pediatrics. 1992;89:1007–9.
159. Blum NJ, Taubman B, Osborne ML. Behavioral characteristics of children with stool toileting refusal. Pediatrics. 1997;99:50–3.
160. Benninga MA, Voskuijl WP, Taminiau JA. Childhood constipation: is there new light in the tunnel? J Pediatr Gastroenterol Nutr. 2004;39:448–64.
161. Voskuijl WP, van Ginkel R, Benninga MA, Hart GA, Taminiau JA, Boeckxstaens GE. New insight into rectal function in pediatric defecation disorders: disturbed rectal compliance is an essential mechanism in pediatric constipation. J Pediatr. 2006;148:62–7.
162. Tagart RE. The anal canal and rectum: their varying relationship and its effect on anal continence. Dis Colon Rectum. 1966;9:449–52.
163. Rao SS, Sadeghi P, Beaty J, Kavlock R, Ackerson K. Ambulatory 24-h colonic manometry in healthy humans. Am J Physiol Gastrointest Liver Physiol. 2001;280:G629.
164. Christensen J. The response of the colon to eating. Am J Clin Nutr. 1985;42:1025–32.
165. Camilleri M, Bharucha AE, di Lorenzo C, et al. American Neurogastroenterology and Motility Society consensus statement on intraluminal measurement of gastrointestinal and colonic motility in clinical practice. Neurogastroenterol Motil. 2008;20:1269–82.
166. Rao SS, Camilleri M, Hasler WL, et al. Evaluation of gastrointestinal transit in clinical practice: position paper of the American and European Neurogastroenterology and Motility Societies. Neurogastroenterol Motil. 2011;23:8–23.
167. Dinning PG, Di Lorenzo C. Colonic dysmotility in constipation. Best Pract Res Clin Gastroenterol. 2011;25:89–101.
168. Martin MJ, Steele SR, Noel JM, Weichmann D, Azarow KS. Total colonic manometry as a guide for surgical management of functional colonic obstruction: preliminary results. J Pediatr Surg. 2001;36:1757–63.
169. Di Lorenzo C, Flores AF, Reddy SN, Hyman PE. Use of colonic manometry to differentiate causes of intractable constipation in children. J Pediatr. 1992;120:690–5.
170. Pensabene L, Youssef NN, Griffiths JM, Di Lorenzo C. Colonic manometry in children with defecatory disorders. Role in diagnosis and management. Am J Gastroenterol. 2003;98(5):1052–7.
171. Dinning PG, Zarate N, Hunt LM, et al. Pancolonic spatiotemporal mapping reveals regional deficiencies in, and disorganization of colonic propagating pressure waves in severe constipation. Neurogastroenterol Motil. 2010;22:e340–9.
172. Dinning PG, Szczesniak MM, Cook IJ. Twenty-four hour spatiotemporal mapping of colonic propagating sequences provides pathophysiological insight into constipation. Neurogastroenterol Motil. 2008;20(9):1017–21.
173. Rao SS, Sadeghi P, Beaty J, Kavlock R. Ambulatory 24-hour colonic manometry in slow-transit constipation. Am J Gastroenterol. 2004;99:2405–16.
174. Herve S, Savoye G, Behbahani A, Leroi AM, Denis P, Ducrotte P. Results of 24-h manometric recording of colonic motor activity with endoluminal instillation of bisacodyl in patients with severe chronic slow transit constipation. Neurogastroenterol Motil. 2004;16:397–402.
175. Leroi AM, Lalaude O, Antonietti M, et al. Prolonged stationary colonic motility recording in seven patients with severe constipation secondary to antidepressants. Neurogastroenterol Motil. 2000;12:149–54.
176. Bassotti G, Gaburri M. Manometric investigation of high-amplitude propagated contractile activity of the human colon. Am J Physiol. 1988;255:G660–4.
177. Brookes SJ, Dinning PG, Gladman MA. Neuroanatomy and physiology of colorectal function and defaecation: from basic science to human clinical studies. Neurogastroenterol Motil. 2009;21:9–19.
178. Garcia-Olmo D, Garcia-Picazo D, Lopez-Fando J. Correlation between pressure changes and solid transport in the human left colon. Int J Colorectal Dis. 1994;9:87–91.
179. Bassotti G, de Roberto G, Castellani D, Sediari L, Morelli A. Normal aspects of colorectal motility and abnormalities in slow transit constipation. World J Gastroenterol. 2005;11:2691–6.
180. Cook IJ, Furukawa Y, Panagopoulos V, Collins PJ, Dent J. Relationships between spatial patterns of colonic pressure and individual movements of content. Am J Physiol Gastrointest Liver Physiol. 2000;278:G329–41.
181. Bassotti G, Iantorno G, Fiorella S, Bustos-Fernandez L, Bilder CR. Colonic motility in man: features in normal subjects and in patients with chronic idiopathic constipation. Am J Gastroenterol. 1999;94:1760–70.
182. Bassotti G, Chistolini F, Nzepa F, Morelli A. Colonic propulsive impairment in intractable slow-transit constipation. Arch Surg. 2003;138:1302–4.
183. Shafik A. Sigmoido-rectal junction reflex: role in the defecation mechanism. Clin Anat. 1996;9:391–4.
184. Scott SM. Manometric techniques for the evaluation of colonic motor activity: current status. Neurogastroenterol Motil. 2003;15:483–513.
185. Bampton PA, Dinning PG, Kennedy ML, Lubowski DZ, deCarle D, Cook IJ. Spatial and temporal

organization of pressure patterns throughout the unprepared colon during spontaneous defecation. Am J Gastroenterol. 2000;95:1027–35.
186. Bassotti G, Gaburri M, Imbimbo BP, et al. Colonic mass movements in idiopathic chronic constipation. Gut. 1988;29:1173–9.
187. Truelove SC. Movements of the large intestine. Physiol Rev. 1966;46:457–512.
188. Halls J. Bowel content shift during normal defaecation. Proc R Soc Med. 1965;58:859–60.
189. Broens PM, Penninckx FM, Lestar B, Kerremans RP. The trigger for rectal filling sensation. Int J Colorectal Dis. 1994;9:1–4.
190. Broens P, Penninckx F. Filling sensations after restorative proctocolectomy. Acta Chir Belg. 2002;102:20–3.
191. Goligher J, Hughes E. Sensibility of the rectum and colon. Its role in the mechanism of anal continence. Lancet. 1951;1:543–7.
192. Lynn PA, Brookes SJ. Function and morphology correlates of rectal nerve mechanoreceptors innervating the guinea pig internal anal sphincter. Neurogastroenterol Motil. 2011;23:88–95.
193. Cook IJ, Talley NJ, Benninga MA, Rao SS, Scott SM. Chronic constipation: overview and challenges. Neurogastroenterol Motil. 2009;21:1–8.
194. Broens P, Vanbeckevoort D, Bellon E, et al. Combined radiologic and manometric study of rectal filling sensation. Dis Colon Rectum. 2002;45:1016–22.
195. Shouler P, Keighley MR. Changes in colorectal function in severe idiopathic chronic constipation. Gastroenterology. 1986;90:414–20.
196. Meunier P, Marechal JM, de Beaujeu MJ. Rectoanal pressures and rectal sensitivity studies in chronic childhood constipation. Gastroenterology. 1979;77:330–6.
197. Baldi F, Ferrarini F, Corinaldesi R, et al. Function of the internal anal sphincter and rectal sensitivity in idiopathic constipation. Digestion. 1982;24:14–22.
198. De Medici A, Badiali D, Corazziari E, Bausano G, Anzini F. Rectal sensitivity in chronic constipation. Dig Dis Sci. 1989;34:747–53.
199. Hoffmann B, Timmcke A, Gathright JJ, Hicks T, Opelka F, Beck D. Fecal seepage and soiling: a problem of rectal sensation. Dis Colon Rectum. 1995;38:746–8.
200. Hancke E, Schurholz M. Impaired rectal sensation in idiopathic faecal incontinence. Int J Colorectal Dis. 1987;2:146–8.
201. Farthing MJ, Lennard-Jones JE. Sensibility of the rectum to distension and the anorectal distension reflex in ulcerative colitis. Gut. 1978;19:64–9.
202. Williams NS, Ogunbiyi OA, Scott SM, Fajobi O, Lunniss PJ. Rectal augmentation and stimulated gracilis anal neosphincter: a new approach in the management of fecal urgency and incontinence. Dis Colon Rectum. 2001;44:192–8.
203. Chan CL, Scott SM, Williams NS, Lunniss PJ. Rectal hypersensitivity worsens stool frequency, urgency, and lifestyle in patients with urge fecal incontinence. Dis Colon Rectum. 2005;48:134–40.
204. Sun WM, Donnelly TC, Read NW. Utility of a combined test of anorectal manometry, electromyography, and sensation in determining the mechanism of 'idiopathic' faecal incontinence. Gut. 1992;33:807–13.
205. Mimura T, Nicholls T, Storrie JB, Kamm MA. Treatment of constipation in adults associated with idiopathic megarectum by behavioural retraining including biofeedback. Colorectal Dis. 2002;4:477–82.
206. Harraf F, Schmulson M, Saba L, et al. Subtypes of constipation predominant irritable bowel syndrome based on rectal perception. Gut. 1998;43:388–94.
207. Di Lorenzo C, Benninga MA. Pathophysiology of pediatric fecal incontinence. Gastroenterology. 2004;126:S33–40.
208. Gattuso JM, Kamm MA, Talbot JC. Pathology of idiopathic megarectum and megacolon. Gut. 1997;41:252–7.
209. Gattuso JM, Smith VV, Kamm MA. Altered contractile proteins and neural innervation in idiopathic megarectum and megacolon. Histopathology. 1998;33:34–8.
210. Ringel Y, Whitehead WE, Toner BB, et al. Sexual and physical abuse are not associated with rectal hypersensitivity in patients with irritable bowel syndrome. Gut. 2004;53:838–42.
211. Lynn PA, Olsson C, Zagorodnyuk V, et al. Rectal intraganglionic laminar endings are transduction sites of extrinsic mechanoreceptors in the guinea pig rectum. Gastroenterology. 2003;125:786–94.
212. Lynn PA, Blackshaw LA. In vitro recordings of afferent fibres with receptive fields in the serosa, muscle and mucosa of rat colon. J Physiol. 1999;518:271–82.
213. Denny-Brown D, Robertson E. An investigation of the nervous control of defaecation. Brain. 1935;58:256–310.
214. White JC, Verlot MG, Ehrentheil O. Neurogenic disturbances of the colon and their investigation by the colonmetrogram: a preliminary report. Ann Surg. 1940;112:1042–57.
215. Lembo T, Munakata J, Mertz H, et al. Evidence for the hypersensitivity of lumbar splanchnic afferents in irritable bowel syndrome. Gastroenterology. 1994;107:1686–96.
216. Chan CL, Facer P, Davis JB, et al. Sensory fibres expressing capsaicin receptor TRPV1 in patients with rectal hypersensitivity and faecal urgency. Lancet. 2003;361:385–91.
217. Gregersen H, Kassab G. Biomechanics of the gastrointestinal tract. Neurogastroenterol Motil. 1996;8:277–97.
218. Gladman MA, Aziz Q, Scott SM, Williams NS, Lunniss PJ. Rectal hyposensitivity: pathophysiological mechanisms. Neurogastroenterol Motil. 2009;21(5):508–16.

219. Petersen P, Gao C, Arendt-Nielsen L, Gregersen H, Drewes AM. Pain intensity and biomechanical responses during ramp-controlled distension of the human rectum. Dig Dis Sci. 2003;48:1310–6.
220. Chan CL, Ponsford S, Scott SM, Swash M, Lunniss PJ. Contribution of the pudendal nerve to sensation of the distal rectum. Br J Surg. 2005;92:859–65.
221. Loening-Baucke V, Yamada T. Is the afferent pathway from the rectum impaired in children with chronic constipation and encopresis? Gastroenterology. 1995;109:397–403.
222. Rao SS, Kuo B, RW MC, et al. Investigation of colonic and whole gut transit with wireless motility capsule and radio-opaque markers in constipation. Clin Gastroenterol Hepatol. 2009;7:537–44. (Epub ahead of print).
223. Gladman MA, Lunniss PJ, Scott SM, Swash M. Rectal hyposensitivity. Am J Gastroenterol. 2006;101:1140–51.
224. Wester C, Brubaker L. Normal pelvic floor physiology. Obstet Gynecol Clin North Am. 1998;25:707–22.
225. Lubowski D, Meagher A, Smart R, Butler S. Scintigraphic assessment of colonic function during defaecation. Int J Colorectal Dis. 1995;10:91–3.
226. Ito T, Sakakibara R, Uchiyama T, Zhi L, Yamamoto T, Hattori T. Videomanometry of the pelvic organs: a comparison of the normal lower urinary and gastrointestinal tracts. Int J Urol. 2006;13:29–35.
227. MacDonald A, Paterson PJ, Baxter JN, Finlay IG. Relationship between intra-abdominal and intrarectal pressure in the proctometrogram. Br J Surg. 1993;80:1070–1.
228. Enck P, Vodusek DB. Electromyography of pelvic floor muscles. J Electromyogr Kinesiol. 2006;16:568–77.
229. Rao SSC, Mudipalli RS, Stessman M, Zimmerman B. Investigation of the utility of colorectal function tests and Rome II criteria in dyssynergic defecation (Anismus). Neurogastroenterol Motil. 2004;16:589–96.
230. Petros PE, Swash M. The musculoelastic theory of anorectal function and dysfunction. Pelviperineology. 2008;27:89–93.
231. Loder PB, Kamm MA, Nicholls RJ, Phillips RK. Haemorrhoids: pathology, pathophysiology and aetiology. Br J Surg. 1994;81:946–54.
232. Lynch AC, Anthony A, Dobbs BR, Frizelle FA. Anorectal physiology following spinal cord injury. Spinal Cord. 2000;38:573–80.
233. Nyam DC. The current understanding of continence and defecation. Singapore Med J. 1998;39:132–6.
234. Vaizey CJ, Kamm MA, Bartram CI. Primary degeneration of the internal anal sphincter as a cause of passive faecal incontinence. Lancet. 1997;349:612–5.

Normal Endosonographic Features of the Anal Canal in Children

5

Alireza Safaei Keshtgar

5.1 Anatomy of Anal Canal and Rectum

The anal canal is the most terminal part of the alimentary tract that extends down from the rectum to open at the anus. The anus lies at the midpoint between the palpable tip of the coccyx and the perineal body. The anal canal remains closed at rest but opens for the passage of faeces or flatus. Its closure is maintained by the tonic rhythmical contraction of the smooth muscle of the internal anal sphincter (IAS) assisted by contractions of the voluntary external anal sphincters (EAS). Anatomical and functional studies of anal canal, including unpublished report in our unit by Lawson, have shown that anal canal in full-term newborn measures 1.5 cm in length and it grows to 2.5 cm by the age of 2 years. There is little further linear growth after this and the effective anal canal measures 2.5–3.0 cm during the rest of the childhood. The anal canal is characterised by 4–6 major folds with intervening minor creases extending up into vertical folds or columns in the lower rectum and down into the folds in closed anus [1–4].

In the anal canal, the dentate line indicates the junction between the superior part of the canal derived from endoderm and the inferior part derived from ectoderm. The nerve supply to the anal canal superior to the dentate line and rectum is visceral innervation by sympathetic and parasympathetic fibres from the inferior hypogastric plexus, which is sensitive to stretch, temperature and electric stimulation. The nerve supply of the anal canal inferior to the dentate line and a narrow band of specialised anal sensation extending about 1 cm above the dentate line are somatic innervations by branches of pudendal nerve, which is sensitive to pain, touch and temperature [5]. The IAS is an involuntary smooth muscle, which is a continuation of circular muscle of rectum surrounding superior two-thirds of anal canal and extending one-third below the dentate line. The EAS sphincter is a voluntary striated muscle surrounding the IAS and inferior one-third of anal canal. The muscle has three components including subcutaneous fibres accounting for anal pucker, superficial fibres providing sphincteric function below the lower limit of the IAS and deep fibres, which blend with the puborectalis muscle. The deep fibres of the external anal sphincter and puborectalis muscle function as a unit. It is known as the striated muscle sphincter complex supplied by sacral nerves from S2 to S4 via pudendal nerve that contains both motor and sensory fibres. The puborectalis is the most anterior and medial part of levator ani muscle that forms a sling posteriorly around the rectum and its contraction pulls the rectum anteriorly forming the anorectal angle of approximately 90°

A. S. Keshtgar (✉)
Evelina London Children's Hospital, Guy's and St Thomas' NHS Foundation Trust, London, UK
e-mail: Ali.Keshtgar@gstt.nhs.uk

M. Lima, G. Ruggeri (eds.), *Ano-Rectal Endosonography and Manometry in Paediatrics*,
https://doi.org/10.1007/978-3-030-97668-2_5

responsible for gross faecal continence [4, 5]. On voluntary contraction, anal canal pressures increase in the proximal part by puborectalis muscle and distal part by the EAS and anorectal angle moves cranially [6].

The primary muscles for faecal continence are the internal and external anal sphincters and the puborectalis muscle. There are controversies regarding the anatomy of the anal canal due to complex overlapping arrangement of the IAS and the EAS and their relations with pelvic floor muscles. In 1715, Santorini described three subdivisions of the EAS muscle as superficial, subcutaneous and deep portion [7] (Santorini, 1715). In 1852, Cruveilhier observed inseparable union between the deep part of the external sphincter and levator ani [8]. In 1897, Holl reported that some of the pubococcygeus muscle fibres loop around the rectum to form a sling defined as puborectalis muscle with no attachment to coccyx [9]. In 1934, Milligan considered puborectalis as part of levator ani muscle and in 1973 OH and colleagues identified deep parts of the EAS and puborectalis as a single muscle [10, 11]. In 1974 Lawson described two distinct groups of the EAS muscles. The deep component above formed a distinct annulus that blended posteriorly with puborectalis and superficial part below, laid medially in the same vertical plane as the IAS and laterally forming a horizontal sheet of muscle bundles in the subcutaneous tissue of the perianal skin [4]. He described that these muscle bundles were divided and anchored by fibres of conjoined longitudinal muscle that spread out distally to insert into the superficial sphincter muscles. However, the sphincter could not be readily separated on gross dissection into a superficial and subcutaneous portion. The upper fibres of deep anal sphincter and puborectalis insert into either side of the perineal body. This has functional significance as contraction of these muscles pulls the pelvic viscera and closes anal canal, vagina and urethra. The endo MRI studies have also shown that the puborectalis is attached to levator ani and deep sphincter. At deep level, the EAS inner fibres encircle the whole circumference of the anal canal and the outer fibres merge with the right and left transverse perineal muscles, where it forms more than 25% of its anterior circumference and joins perineal body. Therefore, transvers perineal muscles contribute to the function of the EAS and damage to these muscles as in obstetric injuries may cause anal sphincter dysfunction [12].

For descriptive purposes, the anal canal can be divided into three parts, upper, middle and lower [4, 13]. The upper part is lined by columnar epithelium that extends down between the anal columns to interdigitate with the processes of stratified cuboidal epithelium. The myenteric plexus extends down in these processes as far as the columnar epithelium and muscularis mucosae. The middle part is lined by stratified cuboidal epithelium. There are thin mucosal folds, forming shallow semilunar valve enclosing six to seven anal sinuses. A series of simple tubular glands open into the sinuses, which extends for a variable distance into the submucosa and IAS.

The lower third is lined with stratified epithelium. Proximally the epithelium is stratified non-squamous and capable of considerable distension. Distally pigmented squamous stratified epithelium lines the canal that extends out through the anus to the perianal skin. This epithelium is hair bearing and characterised by prominent and numerous sebaceous glands. The longitudinal smooth muscle of the rectum extends down into conjoined longitudinal muscle and contributes to the IAS. The conjoined muscle breaks up into fibromuscular septa which insert into the perianal skin between the perianal apocrine glands. These musculotendinous slips pass between the bundles of the subcutaneous part of the EAS and inferior part of the IAS.

5.2 Endosonography Features of Anal Canal

In 1956, Wild et al. described endosonography of the rectum [14]. In 1989, Law and Bartram reported the technique of anal endosonography using type 1846 ultrasonographic scanner and a 7 MHz rotating endoprobe (Bruel & Kjaer, Naerum, Denmark) [13]. They demonstrated sonographic morphology of anal canal in five

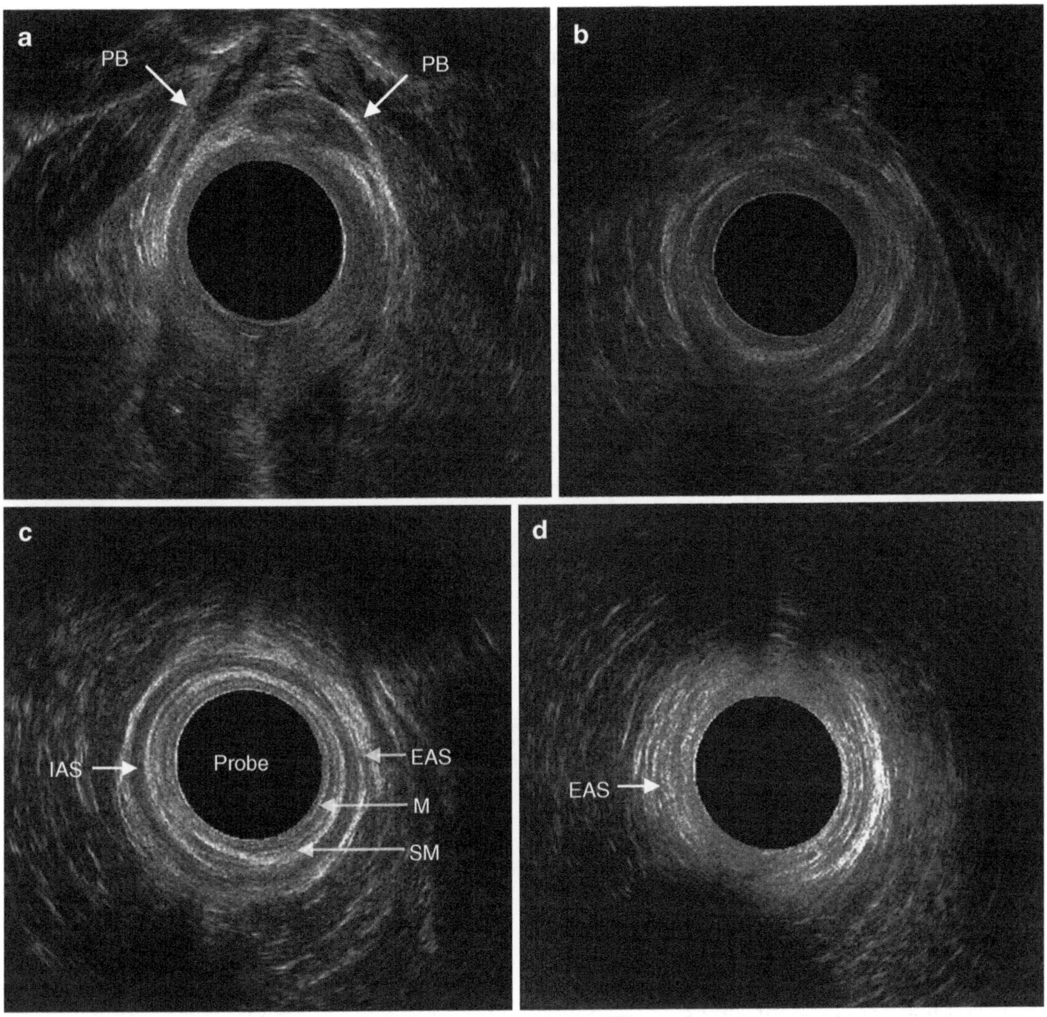

Fig. 5.1 Endosonography scan of normal anal canal in a 9-year-old boy. (**a**) High anal canal, showing hyperechoic U-shaped sling of puborectalis (PB) muscle around lateral and posterior lower rectum with fibres attached to the pubic rami anteriorly at 12 o'clock. (**b**) Upper anal canal, demonstrating the level between the lower border of the PB muscle and the EAS, where it forms a complete ring anteriorly. (**c**) Middle anal canal demonstrating the mucosa as an inner hypoechoic dark ring next to probe, submucosa as hyperechoic bright band, internal anal sphincter (IAS) as a homogenous hypoechoic dark circle and EAS surrounding the IAS as a mixed hyperechogenic bright complete ring. (**d**) Lower anal canal demonstrating the EAS below termination of the internal sphincter

layers including mucosa, submucosa, internal anal sphincter (IAS), intersphincteric plane and external anal sphincter (EAS). The anal endosonography findings have been validated by anorectal manometry, histological examination and electromyography of the sphincters [2, 13, 15].

On endosonography the mucosa and subepithelial tissues appear as an inner hypoechoic dark layer next to the double ring of endoprobe cone (Fig. 5.1b). The submucosa appears as a hyperechoic bright band between the mucosa and the IAS. The circular smooth muscle of the IAS is identified as a homogenous hypoechoic dark circle that surrounds superior two-thirds of the anal canal and extends caudally to just above the anal verge. The longitudinal smooth muscle fibres may also appear surrounding the IAS as a hypoechoic concentric ring with intersphincteric fat in between two layers. The striated muscle of the EAS has a mixed bright hyperechogenic pattern and streaky appearance. The EAS surrounds the IAS and extends from the puborectalis component of the levator ani muscle to its cutaneous termination in inferior one-third of the anal canal. The submucosal vascular plexus, anal cushions and valves cannot be identified due to compression from the endoprobe [13].

The sonographic boundaries of anal canal can be defined as upper, middle and lower levels. At higher level the hyperechoic U-shaped sling of puborectalis (PB) muscle traverses around lateral and posterior part of lower rectum and fibres attach to the pubic rami anteriorly at 12 o'clock position. The EAS is not formed completely at this level (Fig. 5.1a). The upper anal canal is the level between the lower border of the puborectalis muscle and the EAS where it forms a complete ring anteriorly (Fig. 5.1b). The middle anal canal is the level of hyperechoic bright complete ring of the EAS ring, which surrounds the hypoechoic dark circle of the IAS (Fig. 5.1c). The lower anal canal is the level where the IAS is no longer visible and only the EAS and surrounding soft tissues are seen (Fig. 5.1d) [2, 13, 16].

Normal sonography of anal sphincter morphology has shown differences between males and females. In females the EAS is deficient and shorter anteriorly in the region of the perineal body and vagina, whereas in males it is thicker and relatively hypoechoic and narrows anteriorly in the midline. However, the EAS is similar in both sexes posterolaterally. The longitudinal muscle layer can be distinguished sonographically more easily in males, as the external sphincter is relatively hypoechoic [13, 17]. There is no significant correlation between the thickness of the internal and external sphincters and sex in children [18].

5.3 Clinical Application of Anal Endosonography

Endosonography has been used to evaluate the morphology of the anal sphincters and pelvic floor in patients with sphincter defects, anal fistulas and abscesses, and muscular dystrophy [2, 19–22]. The endosonography features of disrupted anal sphincters, in women who have sustained childbirth injuries and in men who have undergone surgery to the anal canal or anal dilatation, have correlated significantly with anal dysfunction [23].

In recent years endosonography has been used more widely in children for the assessment of sphincter deficiencies in patients with constipation and faecal incontinence and injury following repair of anorectal malformation and pull-through surgery for Hirschsprung disease [20, 21, 24, 25]. Endosonography is cheap and effective to assess the structure and integrity of anal sphincters; however it does not distinguish between the smooth muscle of the IAS and the circular muscle of the rectum. Anorectal manometry complements the endosonography to differentiate between the function of the IAS and the rectal muscle by demonstrating rectoanal inhibitory reflex (RAIR). The anorectal manometry also provides additional information about pressure profiles, size and compliance of the rectum, sensation and motor function of the EAS muscle.

The anal endosonography was originally performed using a 2D ultrasound scanner with a 7 or 10 MHz rotating endoprobe providing a 360-degree axial view. The focal point ranged between 5 and 45 mm with better definition of distant structures with higher frequency. The lower frequency of 7 MHz gave a better resolution of near field to identify fistula in ano or pararectal abscess [13]. Three-dimensional (3D) endosonography with 10–16 MHz and high-resolution rotating endoprobe has been used since late 1990. 3D sonography produces digital images that can be seen from different planes permitting measurement of length, width, area and volume with better identification of sphincter defects [26, 27]. The digital volume can be displayed either in multiplanar images of coronal, sagittal and axial in a single image or in tomographic slicing, which allows better visualisation and detection of damage to the anal sphincter muscle complex. There is no difference in endosonography appearance of anal canal in adult and paediatric patients [2, 18].

In children, endosonography is done in the left lateral position using general anaesthesia or sedations, e.g. ketamine, chloral hydrate or awake, depending on the age and cooperation of the child. Most authors have used endoprobe of 10 MHz with hard sonolucent of 17 mm or 8 mm in diameter depending on the size of the anus [2, 18, 28]. The endoprobe is covered with lubricated condom, inserted into the rectum and gently withdrawn from the anal canal caudally by moving the probe or using the built-in mover buttons.

Endosonography features of the anal canal have been investigated using the above technique in normal and constipated children. De la Portilla et al. reported ultrasonography findings of 110 normal children (76 males and 34 females), who were admitted for other reasons like abdominal pain with no history of anorectal problems or surgery (Table 5.1). The mean age was 3.94 years (range 1.1–15.09 years) and mean IAS at middle anal canal was 1.23 mm ± 0.45 (standard deviation) (Table 5.1). There was a significant increase in the thickness of puborectalis and the EAS but not the internal sphincter with age as the IAS thickness remained the same until 10 years of age [18]. Hosie et al. compared 39 normal (control) children with 16 constipated patients and showed a linear correlation between the thickness of the IAS and both age and weight ranging from 0.4 mm in infants to 0.9 mm in adolescents. Children who were constipated had thickening of the IAS ranging from 0.5 to 1.9 mm ($P = 0.005$), which was independent of the duration of symptoms [28]. We reported 92 children (57 males, 35 females), who had symptoms of chronic idiopathic constipation and who were investigated by endosonography, anorectal manometry and colonic transit study (Table 5.1) [25]. The median age was 8.46 years (range 3.35–14.97 years) and median duration of symptoms was 4 years (range 0.3–14.5 years). The median thickness of the IAS was 0.93 mm (range 0.3–2.1 mm) and median functional length of anal canal on manometry was 2 cm (range 1–4 cm). There was a linear correlation between the thickening of the IAS and duration and severity of symptoms, size of megarectum and amplitude of rectal contractions. The study confirmed the earlier findings that the pathogenesis of thickened IAS on endosonography was secondary to the continuous stimulation of the rectum by faecal impaction, resulting in reflex relaxations and contractions of the IAS and hypertrophic changes of the IAS over time [2, 25]. Our results showed that the resting anal sphincter pressure was within normal range and non-obstructive in these patients and there was no justification to do anal dilatation and IAS myectomy for chronic constipation in children. Thickening of the IAS has also been reported in patients with rectal prolapse and solitary rectal ulcer syndrome (SRUS) and the underlying pathophysiology is analogous to chronic constipation in children [29, 30].

On endosonography, defects in the IAS or EAS appear as amorphous areas of varying echogenicity and are seen as breaks in the continuity of the muscle ring. A defect in the IAS appears as a hyperechogenic area and disruption of the EAS appears as a hypoechoic area [31, 32]. In patients

Table 5.1 Endosonography findings of middle anal canal in normal and constipated children

	Normal ($n = 110$) [18]	Normal ($n = 39$) [28]	Constipation ($n = 16$) [28]	Constipation ($n = 92$) [25]
Age (year)	3.94 (1.10–15.09)[a]	Infant to adolescent	6 months to 13 years	8.46 (3.35–14.97)[b]
Internal sphincter	1.23 mm ± 0.45 (SD)[c]	(0.4–0.9 mm)[b]	(0.5–1.9 mm)[b]	0.93 mm (0.3–2.1)[b]
External sphincter	4.47 mm (3.70–5.70)[d]			

[a] Mean (range)
[b] Denotes median (range)
[c] Denotes mean standard deviation (SD)
[d] Denotes median (range P_{25} to P_{75})

with incomplete defects or atrophy, relative thinning rather than actual loss of the continuity of the muscle ring is seen. Lateral internal anal sphincterotomy or myectomy creates a well-defined defect with rounded ends involving upper half or lower third of the IAS [17, 31]. The remaining sphincter bunches up to look thicker because the cut ends spring apart [33]. The scar tissue in the sphincter appears as sonographic areas of mixed echogenicity [34].

In adults the lesions are classified according to Starck criteria to define the degree of obstetric sphincter injury [35]. The defect is scored according to the length (≤½, >½ and whole circumference) of the sphincter, depth (partial or total) and size ≤90°, 91°–180° and >180°. The score ranges from 0 to 16 and the sphincter damage is classified as small (score 1–4), moderate (score 5–7) or large (score 8–16). We have devised a simplified endosonography score based on the integrity of the IAS muscle in children (Table 5.2) [20]. The score ranges from 0 to 3 and correlates well with functional anorectal manometry score and modified Wingfield score to assess faecal continence with bowel function.

Endosonography provides valuable information related to the integrity of anal sphincter muscles following repair of anorectal malformation and Hirschsprung disease and assessment of faecal incontinence. Evidence from the literature and our experience indicate that endosonography complements the anorectal manometry study and both are instrumental to evaluate the morphology and function of anal sphincters and outcome of faecal continence. These studies help to understand the causes of dysfunction and develop clinical care pathway to guide the management in infancy and preschool years so that appropriate measures can be implemented at an early stage. This is discussed further in Chap. 13.

Table 5.2 Endosonography scoring system for internal anal sphincter integrity in children [20]

Internal anal sphincter	Score[a]
Normal	0
Scarred	1
Fragmented	2
Absent	3

[a] Scores of 0–1 indicate good and scores of 2–3 indicate poor-quality IAS

References

1. El-Shabrawi MHHM, Abdelgawad MMAH, Hassanin F, Mahfouze AAA, Khalil AFMESE. High-resolution anorectal manometry in children with functional constipation: a single-Centre experience before and after treatment. Gastroenterology Rev. 2018;13(4):305–12.
2. Keshtgar AS, Ward HC, Clayden GS, Sanei A. Thickening of the internal anal sphincter in idiopathic constipation in children. Pediatr Surg Int. 2004;20(11–12):817–23.
3. Kumar S, Ramadan S, Gupta V, Helmy S, Atta I, Alkholy A. Manometric tests of anorectal function in 90 healthy children: a clinical study from Kuwait. J Pediatr Surg. 2009;44(9):1786–90.
4. Lawson JON. Pelvic anatomy II. Anal canal and associated sphincters. Ann R Coll Surg Engl. 1974;54:288–300.
5. O'Neill JA Jr, Grosfeld JL, Fonkalsrud EW, Coran AG. Other disorders of the anus and rectum, anorectal function. In: Craven L, editor. Pediatric surgery. 2. 5th ed. St Louis, Missouri: Mosby; 1998. p. 1449–60.
6. Raizada V, Bhargava V, Karsten A, Mittal RK. Functional morphology of anal sphincter complex unveiled by high definition anal manometery and three dimensional ultrasound imaging. Neurogastroenterol Motil. 2011;23(11):1013–9. e460.
7. Santorini GD. Anatomici summi septem decim tabulae. Parmac, ex Regiat Topographia. 1715;(VI and VII). I775 edn.
8. Cruveilhier J. Traiti d' Anatomie descriptive. 3rd ed. Paris: Labe; 1852. p. 662.
9. Holl M. Handbuch der Anatomie, ed: Bardelben. Jena: Fischer; 1897.
10. Milligan ETC, Morgan CN. Lancet. 1934;2:1150–213.
11. Oh C, Kark AE. Anatomy of the perineal body. Dis Colon Rectum. 1973;16(6):444–54.
12. Hussain SM, Stoker J, Lameris JS. Anal sphincter complex: endoanal MR imaging of normal anatomy. Radiology. 1995;197(3):671–7.
13. Law PJ, Bartram CI. Anal endosonography: technique and normal anatomy. Gastrointest Radiol. 1989;14(4):349–53.
14. Wild JJ, Reid JM. Diagnostic use of ultrasound. Br J Phys Med. 1956;19(11):248–57; passim.
15. Law PJ, Kamm MA, Bartram CI. A comparison between electromyography and anal endosonography in mapping external anal sphincter defects. Dis Colon Rectum. 1990;33(5):370–3.
16. Gold DM, Halligan S, Kmiot WA, Bartram CI. Intraobserver and interobserver agreement in anal endosonography. Br J Surg. 1999;86(3):371–5.
17. Sultan AH, Kamm MA, Nicholls RJ, Bartram CI. Prospective study of the extent of internal anal

sphincter division during lateral sphincterotomy. Dis Colon Rectum. 1994;37(10):1031–3.
18. de la Portilla F, Lopez-Alonso M. Endosonography of the anal canal: findings in children. Dis Colon Rectum. 2009;52(4):711–4.
19. Eckardt VF, Nix W. The anal sphincter in patients with myotonic muscular dystrophy. Gastroenterology. 1991;100(2):424–30.
20. Keshtgar AS, Athanasakos E, Clayden GS, Ward HC. Evaluation of outcome of anorectal anomaly in childhood: the role of anorectal manometry and endosonography. Pediatr Surg Int. 2008;24(8):885–92.
21. Keshtgar AS, Ward HC, Clayden GS, de Sousa NM. Investigations for incontinence and constipation after surgery for Hirschsprung's disease in children. Pediatr Surg Int. 2003;19(1–2):4–8.
22. Law PJ, Talbot RW, Bartram CI, Northover JM. Anal endosonography in the evaluation of perianal sepsis and fistula in ano. Br J Surg. 1989;76(7):752–5.
23. Lunniss PJ, Gladman MA, Hetzer FH, Williams NS, Scott SM. Risk factors in acquired faecal incontinence. J R Soc Med. 2004;97(3):111–6.
24. Emblem R, Morkrid L, Bjornland K. Anal endosonography is useful for postoperative assessment of anorectal malformations. J Pediatr Surg. 2007;42(9):1549–54.
25. Keshtgar AS, Ward HC, Clayden GS. Pathophysiology of chronic childhood constipation: functional and morphological evaluation by anorectal manometry and endosonography and colonic transit study. J Pediatr Surg. 2013;48(4):806–12.
26. Christensen AF, Nyhuus B, Nielsen MB, Christensen H. Three-dimensional anal endosonography may improve diagnostic confidence of detecting damage to the anal sphincter complex. Br J Radiol. 2005;78(928):308–11.
27. Wisser J, Schar G, Kurmanavicius J, Huch R, Huch A. Use of 3D ultrasound as a new approach to assess obstetrical trauma to the pelvic floor. Ultraschall Med. 1999;20(1):15–8.
28. Hosie GP, Spitz L. Idiopathic constipation in childhood is associated with thickening of the internal anal sphincter. J Pediatr Surg. 1997;32(7):1041–3. discussion 3–4
29. Keshtgar AS. Solitary rectal ulcer syndrome in children. Eur J Gastroenterol Hepatol. 2008;20(2):89–92.
30. Marshall M, Halligan S, Fotheringham T, Bartram C, Nicholls RJ. Predictive value of internal anal sphincter thickness for diagnosis of rectal intussusception in patients with solitary rectal ulcer syndrome. Br J Surg. 2002;89(10):1281–5.
31. Keshtgar AS, Ward HC, Sanei A, Clayden GS. Botulinum toxin, a new treatment modality for chronic idiopathic constipation in children: long-term follow-up of a double-blind randomized trial. J Pediatr Surg. 2007;42(4):672–80.
32. Tjandra JJ, Milsom JW, Schroeder T, Fazio VW. Endoluminal ultrasound is preferable to electromyography in mapping anal sphincteric defects. Dis Colon Rectum. 1993;36(7):689–92.
33. Bartram CI, Frudinger A. Handbook of anal endosonography. Petersfield, UK; Bristol, PA: Wrightson Biomedical Pub; 1997.
34. Engin G. Endosonographic imaging of anorectal diseases. J Ultrasound Med. 2006;25(1):57–73.
35. Starck M, Bohe M, Valentin L. Results of endosonographic imaging of the anal sphincter 2-7 days after primary repair of third- or fourth-degree obstetric sphincter tears. Ultrasound Obstet Gynecol. 2003;22(6):609–15.

6 Anorectal Manometry, Conventional and High Resolution in Paediatrics

Eleni Athanasakos and Stewart Cleeve

6.1 Introduction

Defaecatory disorders (DD) in children and young adults, including chronic constipation (CC) and faecal incontinence (FI), are common conditions worldwide and have a significant impact on children, their families and the healthcare system. CC is an important health problem worldwide in children, with a prevalence range of 0.7–29.6% occurring in all age groups, from newborns to young adults [1–5]. The wide range in stated prevalence could be attributed to differences in the definition of constipation (such as cultural understanding of normal bowel movement).

In over 90% of children who present with CC, there is no known organic reason (such as Hirschsprung disease [HSCR] or anorectal anomalies [ARAs]) and these children are considered to suffer from functional constipation (FC) [6]. There is no unanimous international agreement on the definition or classifications of constipation in children within the literature [1, 7] which could be due to the lack of pathophysiological understanding and its multifactorial nature of the condition [2, 8, 9]. Yet, FC is commonly defined as difficulty or delayed defaecation (two or less per week), significant enough to cause distress, and is often associated with FI, faecal impaction, poor appetite, abdominal and anorectal pain, overflow incontinence, rectal bleeding and stool withholding with symptoms present for longer than 4 weeks [10]. Diagnosis is based commonly on Rome IV Criteria [10]. However, diagnosis can be complicated and difficult due to the fact that in children diagnosis is often delayed or implied to be a behavioural issue, and relies on parents' interpretation and reports of the symptoms, multiple efforts and trial and error of medication and other treatments [11, 12] and lack of diagnostic tools and national consensus.

It has gradually become accepted that CC and FI are the results of a pathological storm—that is, it affects both the physical and psychological elements. FI is defined as the passage of a stool without control (usually in a socially inappropriate place) after 4 years of age, at least once a month over a 3-month period [1–4, 13–19]. When first presented, FI may be associated with CC, which causes misguided management, yet studies have shown that over 80% of children with CC also suffer from FI [20, 21] and thus their coexistence is now universally accepted. FI has an estimated worldwide prevalence in children of up to 8% with the majority of children having it secondary to CC [1, 13]. However, up to 5% of

E. Athanasakos (✉) · S. Cleeve
Department of Paediatric Surgery, Level 8, Paediatric Offices, The Royal London Hospital, Barts Health, Whitechapel, UK
e-mail: eleni.athanasakos@nhs.net; stewartjames.cleeve@nhs.net

M. Lima, G. Ruggeri (eds.), *Ano-Rectal Endosonography and Manometry in Paediatrics*,
https://doi.org/10.1007/978-3-030-97668-2_6

patients are affected by other medical conditions that lead to delayed transit or sphincter dysfunction such as HSCR [22, 23], spina bifida [24] or ARAs [1, 13] and 20% are associated with conditions such as non-retentive FI [18], or pelvic floor dyssynergia [25].

Generally, CC/FI is managed in outpatient and community clinics, yet there are increasing numbers of patients presenting to emergency departments [14]. FI is particularly distressing and is associated with poor quality of life (QoL), reduced school attendance and social interaction [26, 27]. Children with CC/FI report a worse QoL compared with children affected by inflammatory bowel disease and gastro-oesophageal reflux [28]. Most of the published guidelines are directed toward the management of FC and provide a framework and general guidance [29, 30]. It is expected practice that adherence to NICE Clinical Guideline [15] will be sufficient to diagnose and successfully conservatively treat (i.e. laxatives, toilet training, disimpaction, maintenance therapy and long-term follow-up) the majority of children, with specific focus on FC. However, some patients fail to respond to conservative measures and continue to live with a disabling condition and a poor quality of life [1–5, 7, 14, 15, 28]. Furthermore, there remains wide variation in the use of diagnostic tools and management strategies for children with DD [1, 5, 29, 31].

The underlying aetiology and pathophysiology of FI and defaecation disorders in children remain multifactorial. Despite the fact that studies have shown that clinical examination and experience alone are ways to both diagnose and treat these patients [32, 33], the rise of recent advances in diagnostic investigations shows otherwise, which demonstrated that a symptom-based assessment seems unsatisfactory to direct treatment modalities [33–36]. The use of anorectal manometry (ARM) to study the physiology of the anorectum is a valuable tool to understand the underlying pathophysiology of DD in children and furthermore direct treatments to best achieve satisfactory outcomes and alleviate the distress of this debilitating condition.

This chapter specifically discusses the role of ARM within paediatrics.

6.2 Definition of Anorectal Manometry (ARM)

The concept of performing anorectal manometry (ARM) to gain pathophysiological information about the anorectum was firstly acknowledged by Gowers in 1887 [37]. Gowers (1887) demonstrated how to measure anal canal resting tone and recto-inhibitory reflex (RAIR). ARM is a remarkable instrument, which provides an objective measurement of anal sphincteric pressures and sensorimotor function of the anorectum in regard to both mechanisms of continence and defaecation [38–41]. ARM has been considered the gold standard tool (in adults specifically) for the assessment of functional dysfunction in anal sphincter tone, presence of the RAIR and alteration of rectal sensory function [40, 42, 43].

ARM is performed by simply placing a specifically designed catheter with a small balloon into the lower rectum and anal canal. The catheter is pressure sensitive and is connected to a transducer, through which mechanical signals are altered into electronic signals that are recorded and displayed on a monitor. It measures the following: high-pressure zone (which refers to the length of the anal sphincter muscles); involuntary function of the anal canal at rest; voluntary anal function on squeezing; RAIR; rectal sensitivity and compliance; rectoanal coordination during simulated defaecation ('push'); and capacity to expel a balloon.

6.3 Anorectal Manometry in Paediatrics

In children with DD, the use of ARM is a specialised investigation to fully understand the pathophysiological mechanisms involved [8], direct management and alleviate the distress of the patients' symptoms. ARM has been considered the gold standard tool for the assessment of anorectal function in adults for over a century [40] with high-resolution ARM (HR-ARM) gaining momentum [44]. In adult practice, ARM is used to guide the management of DD [45–47]. In paediatrics, however, ARM has yet to be fully implemented as part of an accepted routine investigation, in children presenting with similar symptoms. The delayed and to some extent neglected implementation within paediatrics is due to the lack of standardisation for its use, indications (e.g. age, which patients would benefit, diagnostic/prognostic purpose), protocol (awake or under general anaesthesia [GA]/sedation) and type of equipment (traditional versus HR-ARM, water/solid state). To address these problems, a recent national consensus was recently published and reached to utilise its use in the UK [48] and will be referenced throughout this chapter as a useful guideline.

The use of ARM in the awake state or under general anaesthesia/sedation is an area of great controversy. Primarily in the UK, ARM testing in children has historically been performed under sedation [49, 50], because of the child's age, due to the likelihood that they will be uncooperative or simply because ARM has been perceived as an invasive procedure [51]. Traditional ARM in an awake child has been primarily used in isolated research cases (e.g. HSCR) or specific conditions [52] with limited use of the novel HR-ARM [53]. Performing ARM under sedation provides limited physiological parameters due to the discrepancy with type of anaesthesia and level of sedation used that can interfere with the physiological outcomes [54].

Currently, the literature demonstrates a wide variation in the use of ARM for children with DD. The ANMS-NASPGHAN document [51] was the first attempt to offer opinions about the standardisation of ARM practice within paediatrics and provided a platform for the most recent and updated consensus published by the British Society of Paediatric Gastroenterology, Hepatology and Nutrition (BSPGHAN)-Motility Working Group [48].

The aim of this section is to discuss the use of ARM within paediatrics, with specific focus on the indication of use, protocol and existing variability we are currently facing.

6.3.1 Demographic Characteristics

A question that is often asked within clinical practice is: 'At what age can or should you perform ARM'? There has been a male predominance found in the majority of papers (62%), according to a recent systematic review [55]. Age, at which ARM is performed, is largely variable: birth to 18 years old, demonstrating a variable age group. Most of the published literature includes a range that covered multiple age groups; however most common age groups were most represented in young children aged 2–12 with less representation from adolescents aged 12–18 years, infants aged 1–24 months and neonates 0–30 days.

6.3.2 Clinical Indications

The role of manometric examination allows recognising a multitude of underlying deficits that are frequently seen in children with DD. ARM is used in children, just as in adults, to measure sphincter function RAIR, anorectal coordination and rectal sensation. Furthermore, ARM is also useful in identifying whether children can differentiate between a squeeze and push and recognise both their endurance squeeze to prevent FI and their ability to understanding when they need to defaecate.

Indications for the use of ARM within paediatrics remain variable, with limited guidance as to when it should be performed [48]. ARM has been commonly performed in patients with organic conditions (e.g. HSCR) and less commonly in FC and for the assessment of outcomes after surgeries and non-treatments, and rarely was it used for diagnostic (e.g. mainly for HSCR) or research purposes [55]. Based on the recent BSPGHAN consensus, a summary of indications is demonstrated in Table 6.1 [48].

6.3.2.1 Faecal Incontinence and ARM

As discussed, FI is defined as the inability to control the discharge of liquid/solid/gas which can be attributed to various physiological mechanisms of continence. Impairment of anorectal

Table 6.1 Clinical indications and interpretation of ARM/HRAM parameters [48]

Parameter	Rationale	Outcome and clinical implications
Resting pressure	Assessment of anal sphincter baseline integrity	**High:** Muscle spasm (voluntary or involuntary) Functional contraction, e.g. related to anxiety and pain (such as in anal fissure) Anal stenosis/stricture (should be assessed for by gentle digital rectal examination pre-procedure following appropriate consent) **Low:** Weak/hypotensive anal sphincter Idiopathic Drug induced (sedation, anaesthetic) Injury (trauma, abuse) Neurological (spinal cord disorder)
Squeeze and endurance squeeze	Assessment of anal sphincter contractile integrity	**Low maximum pressure:** Non-compliant/poor understanding Disorder of the anal sphincter (neurogenic or myogenic) Injury **Reduced endurance pressure:** Non-compliant child/poor understanding Nerve damage
Push	Assessment of coordination (in conjunction with anal sphincter pressure (ASP))	**Adequate pressure with high ASP** Type 1 dyssynergia **Poor push with high ASP** Type 2 dyssynergia **Adequate pressure with no decrease in ASP** Type 3 dyssynergia **Poor push with no decrease in ASP** Type 4 dyssynergia
Cough	Assessment of sacral reflex arc	**Impaired response** Suggestive of damage to sacral reflex arc
RAIR	Functional assessment of presence of endogenous anorectal neural network	**Positive RAIR:** Excludes Hirschsprung disease **Negative RAIR:** Possible Hirschsprung disease (aganglionic rectal biopsy) Anal sphincter achalasia (ganglionic rectal biopsy) Partial RAIR has been suggested in anorectal inflammatory conditions (e.g. allergy)
Rectal sensation	Assessment of rectal sensation	Educate children in appreciating the different sensations involved in continence and defaecation General marker for rectal capacity and compliance, e.g. in children with FC Impaired sensation may be seen in neurological disturbances such as spinal cord disorders

function can be due to sphincteric and/or sensorimotor or behavioural or psychological triggers. Patients with FI often present with symptoms of urgency, withholding or difficulty in holding the stool in case of urgency [56–58].

The use of ARM in patients with FI is somewhat essential. Dysfunction of the internal anal sphincter (IAS) results in decreased anorectal pressure, which has been associated with passive FI: 'not [being] aware of defaecation' [59–62]. Dysfunction of the external anal sphincter (EAS) has been shown to lead to decreased squeeze anorectal pressures, which are associated with urge FI: 'the inability to hold on' [61, 63–65]. Individuals may present with bowel symptoms—including FI—even if they have an anatomically intact and normal-functioning sphincter complex. This highlights the importance of other pathophysiological mechanisms, such as rectal sensation that needs to be taken into consideration in the ARM protocol [34, 35, 49, 65–69].

Theories have been postulated what the ARM findings mean, in response to DD. Kesthgar et al. (2013) [70] found that patients with CC who have FI were confirmed to have reduced anal sphincter resting tone due to reflex relaxation of the IAS in response to faecal retention in the rectum without the perception of the need to defaecate. Thus, children with faecal impaction presented with higher rectal pressure and lower anal resting sphincter pressure compared to children without faecal impaction. This emphasises the importance of rectum in the pathophysiology of constipation and FI [71].

Disruption of the afferent nerve pathway alters sensory perception, and this deficit has been associated with symptoms such as FI [66, 72]. The afferent pathway can be altered in multiple ways in patients. An *elevated* sensory threshold will result in 'rectal *hypo*sensitivity' and conversely, a *decreased* sensory threshold will result in 'rectal *hyper*sensitivity'. Rectal hyposensitivity has been found in patients with FC [72]. Rectal hypersensitivity has been linked to autonomic neuropathy, congenital neurogenic anorectal malformation, spinal bifida, myelomeningocele, HSCR and functional and somatic alterations of the rectal reservoirs, such as megarectum and descending perineum syndrome. Disturbed sensory function has been reported in patients with a megarectum [66, 72] and in children with FC [73]. Fathy et al. (2013) [74] demonstrated an increase in the maximum tolerable capacity in children with CC compared to healthy children (i.e. without CC). This was confirmed by Li et al. (2008) [161], who found a significant decrease in the sensitivity of the rectum in children with FC with a remarkable increase in maximal tolerated volume. Based on these studies, it was suggested that patients with CC have a lost response to normal stool volume, and retain faecal stools in the rectum, causing evacuation difficulties, when a response aroused. There remains limited literature regarding the use of ARM in patients with functional non-retentive FI [7, 12, 17, 73, 75–78] and two studies included patients with functional abdominal pain [10, 73].

6.3.2.2 Constipation

According to a recent systematic review [55], it has been found that some papers have discussed the use of ARM for FC, for diagnostic or management purposes. Authors typically excluded the following factors in making this diagnosis: anorectal malformation, anal sphincter damage, endocrine disease, metabolic disease, spinal abnormality, learning difficulties and a history of recent surgery or use of drugs known to cause constipation.

The use of both ARM and endosonography has been provided in the literature, some insights of the underlying pathophysiology and structural abnormalities that are seen in children with DD. These include large-size rectum, lower contractility of rectum, impaired rectal sensation, hypertonic/hypotonic or a paradoxical effect of the anal sphincter pressures, decreased relaxation of the IAS in response to rectal distension and thickening of the IAS [79–83]. To date, it remains unclear whether such ARM findings are the primary cause or the secondary effect of CC/FI. There have been conflicting views regarding CC and what is found in ARM regarding the IAS. Some studies have found no significant difference in the anal sphincter resting tone between constipated and healthy children [74, 84, 85]. Yet, other studies have found a significantly higher [18, 86, 87] or lower resting anal sphincter tone in children with CC [88].

Impaired sensation has been another physiological parameter of significance when understanding CC. Studies have shown that the threshold for first sensation and desire to defaecate can be higher in 60% of patients with DD and are associated with an impaired rectal sensation, generally an increased rectal compliance, with rectal hyposensitivity, indicative of an excessively lax rectum; higher volumes of rectal distension are required to elicit perception also in patients with important dilatation of the rectum (megarectum) as frequently seen in children [89, 90].

It has been found that central brain activity and perception of visceral stimuli can be triggered by emotions, attention and cognitions [91]. This can be applied to within paediatrics for example. It is well recognised that many children do not respond to the urge to defaecate while they are distracted by their games/toys, etc. Furthermore, it is clear that the majority of children develop FC due to stool-withholding behaviour. As a consequence of painful defaecation, many children stiffen their legs and contract their pelvic floor muscles, thereby counteracting propulsive recto-colonic activity. The response to ignore the urge to defaecate may be a conscious decision of the child or an automatism (an unconscious habit), resulting from altered or diminished brain processing of urge sensations due to loss of attention. Constipation due to stool withholding can generally be successfully treated with laxative therapy in combination with behavioural modifications. However, despite intensive therapy, a subgroup of patients remain symptomatic up to adolescence or young adulthood and experience many relapse episodes [92]. These patients are problematic because constipation and symptoms such as abdominal pain and associated FI may dramatically influence the quality of life.

There has been evidence that both pelvic floor dysfunction and delayed colonic transit are demonstrable as independent or combined triggers in adolescents with refractory constipation [93]. The question of whether there is correlation with symptoms of constipation and ARM has been addressed in the literature, with conflicting outcomes. Borowitz et al. (1996) [94] demonstrated that the presence of sphincter spasm correlated with the frequency of FI, age of onset and duration of symptoms. Yet Feinberg et al. (2008) [95] did not find correlation with FI and presence of paradoxical puborectalis contraction. Instead, a large proportion of patients had 'inappropriate squeezing when asked to strain' (i.e. 90.5% of patients) [95]. A well-known explanation for this is that children with CC with or without FI have a clinical history of painful defaecation that leads to withholding [94, 96]. In fact, paradoxical puborectalis contraction may be an involuntary form of withholding and could be an integral component in the pathophysiology of CC [95]. Furthermore, Feinberg et al. (2008) [95] demonstrated other ARM parameters which correlated with the symptoms of CC: (i) there was frequency of FI and the volume of first rectal urge; (ii) frequency of encopresis and higher volumes required to elicit the RAIR; and (iii) presence of withholding behaviour and maximum volume tolerated. Thus, FI may be a trigger when performing ARM that higher balloon volumes are needed to elicit the RAIR. This highlights the importance to perform ARM awake in order to incorporate multiple physiological parameters when managing patients with CC. We are still at the infancy of ARM within paediatrics, with limited pathophysiological understanding, and whether such findings are a result of chronic distension or whether it is present before developing constipation is not known.

6.3.2.3 Other

ARM has been used in the diagnosis and efficacy of post-surgery for HSCR for over a century which has been demonstrated in the depth of literature found in this area [52, 97–130]. Chung et al. (2015) [99] demonstrated the benefits of using ARM to assess patients post-operatively after primary transanal endorectal pull-through for HSCR, with better understanding of clinical outcomes, thus intervening early to not only improve bowel dysfunction, but also enable patients to have improved QoL.

This is also the case in children with ARAs, who also need an understanding of the anorectal functional outcome, after reconstructive surgery. It is appreciated that voluntary bowel control is frequently poor after surgery for ARAs [131], with high rates of FI and constipation after all grades of reconstruction [132–134]. Kumar et al. (2010) [135] found on ARM short anal canal with lower resting pressure and impaired RAIR in patients with ARAs. Athanasakos et al. (2008) [8] investigated using ARM patients with all types of ARAs after different reconstructive surgical approaches. Various pathophysiologies were identified in these patients, who presented primarily with FI. Structural integrity of the anal sphincters was a major factor found, but extrasphincteric mechanisms, notably rectal sensory function, were considered equally important. Although there was a tendency for both clinical and pathophysiological differences to exist between subgroups (for example more patients with high anomalies appeared to have a hypersensitive rectum, and a greater proportion of patients with intermediate anomalies had constipation), it was impossible to draw firm conclusions.

Indication for ARM has also been used to evaluate non-surgical treatments. Examples of the latter include cisapride [22], topical isosorbide dinitrate [112, 116] and a bowel management program (Lombardi et al.). Keshtgar et al. (2007) [136] compared outcomes following a surgical approach (internal anal sphincter myectomy) to those of a non-surgical approach (injection of intersphincteric botulinum toxin) in the treatment of FC using ARM. Few studies have used ARM to assess the efficacy of biofeedback therapy [73, 84, 137–147]. Caruso et al. (2015) [148] demonstrated in patients with ARAs who have symptoms of FI the benefit of biofeedback therapy which was supported scientifically with the use of ARM, with potential sphincter recovery after therapy seen, which was also correlated to morphologic evaluation with MRI. It has been demonstrated using ARM for biofeedback therapy that abnormal defaecation dynamics has been found in 50% of children with FC [84, 87] and rectal barostat studies demonstrate impaired rectal sensation and higher rectal compliance [149]. Biofeedback therapy may be an option in children with chronic DD, alongside conservative treatment, yet needs to be further explored.

Limited studies have used ARM in the context of neurological origin including patients with spina bifida [75, 105, 129, 130, 150], tethered cord syndrome [151–153] and neuronal intestinal dysplasia [154, 155]. ARM has also been used for research purposes including sphincteric dysfunction associated with constipation, incontinence and other paediatric conditions [70, 74]; the effects of anaesthetic agents on the RAIR [54, 156]; and the contribution of ARM to other imaging modalities such as MRI in the identification of spinal cord lesions in children [151]. Spinal cord abnormalities have been shown to increase the tone of the anal sphincters as a result of the damage of upper motoneurons and an exaggerated contraction and 'anal spasms' upon balloon dilation or sphincter relaxation with smaller balloon inflating volumes [151]. Equally, in some neurological conditions, it has been shown that the anal tone may be decreased due to lower motoneuron deficit [157].

6.3.3 ARM Parameters

The ARM parameters assessed most frequently are the RAIR and anal resting pressure, which is of no surprise, as ARM has been conventionally done under general anaesthesia (GA), which will be discussed later [55]. Lesser emphasis has been based on the other parameters such as squeeze pressure/increment with minimal description on the protocol and analysis of how this is measured and even fever studies to include physiological parameters such as maximum squeeze pressures [138, 158], cough reflex [52, 114, 159, 160] and simulated defaecation which is also known as 'push' or 'defaecation dynamics'. Rectal sensation has to some extent been explored, which mostly included studies using progressive balloon inflation to assess minimal pressure thresholds for sensation, urge to defaecate and maximum tolerable volume [53, 63, 74, 86, 94, 105, 161–163]. Based on the recent BSPGHAN consensus [48], a terminology table (Table 6.2) provides a quantitative assessment and a qualitative assessment of the other parameters.

6.3.4 Use of Sedation or General Anaesthesia (GA)

ARM has been historically performed under sedation or GA in children [49, 50], because the child is too young or uncooperative, or because ARM has been perceived as an invasive procedure [51]. In a recent systematic review [55], the existing vast variation in the use of GA in ARM was demonstrated. Studies have demonstrated

Table 6.2 High-resolution anorectal manometry paediatric parameters [44, 46, 48]

Manoeuvre	Definition
FUNCTIONAL ANAL CANAL LENGTH (FACL)	Length of anal canal (cm) in which pressure exceeded rectal pressure by >5 mmHg
Average anal resting pressure	Average maximum pressure (mmHg) over the FACL during the 30-s period of rest
Maximum incremental anal squeeze pressure	Maximum recorded pressure (mmHg) at any point during voluntary squeeze, minus the mean maximum resting pressure prior to the manoeuvre (over 5 s)
Average anal squeeze pressure	Mean maximum pressure (mmHg) sustained over the duration of the 5-s squeeze manoeuvre minus the mean maximum resting pressure prior to the manoeuvre (over 5 s)
Endurance squeeze duration	Length in time (15 s) over which a pressure at or above 50% of the highest recorded squeeze pressure was sustained. The endpoint was determined as the point at which the pressure first dropped below this threshold
Anal cough pressure	Highest recorded pressure within the anal canal (mmHg) at any point during the cough manoeuvre, minus the maximum resting pressure prior to the manoeuvre (over 5 s)
Push: used to assess defaecation dynamics	Qualitatively analysed: The use of the colour contour plots clearly highlighting co-ordinated rectoanal events during the push manoeuvre

that GA agents (nitrous oxide, halothane, althesin, phenoperidine, droperidol, propofol and trimeprazine) inhibit the IAS and reduce resting pressure, yet they seem to have no effect on the qualitative relaxation of the anal sphincter in response to rectal distension as you do when examining the RAIR [95, 128, 164, 165]. This effect has also been demonstrated after administration of inhalational anaesthesia agents such as sodium pentothal or nitrous oxide [166]. However, Keshtgar et al. (2015) [50] found no significant differences in resting pressure results, when using ketamine or awake ARM. Ketamine anaesthesia does not affect quantitative or qualitative measurements of autonomic anorectal function and can be used reliably in children who will not tolerate the manometry while awake [50]. Paradoxical contraction of the EAS can only be evaluated in the awake children and should be investigated further as the underlying cause of obstructive defaecation in patients with DD.

Half of the published studies perform ARM awake [55]; yet not all physiological parameters are acquired during conscious, voluntary anorectal activity. The ARM parameters assessed most frequently in this review were the RAIR and anal resting pressure, with most studies excluding superior physiological parameters (e.g. squeeze increments, push, rectal sensation) that the adult practice uses routinely to guide management in such patients [40].

Based on the recent BSPGHAN consensus, performing ARM under sedation provides limited physiological measurements as the discrepancy with type of anaesthesia and level of sedation used that can interfere with the physiological outcomes [167, 168]. If the ARM procedure is carried out awake, rather than under sedation, more physiological parameters can be obtained (Table 6.3). It is, however, recognised that in certain circumstances (e.g. significant child distress or anxiety, learning difficulties or in infants with suspected HSCR) ARM can be done under sedation. If sedation is needed, muscle relaxants should be avoided and clarity given to the parents and recipient healthcare professionals with regard to the limited information obtained (essentially limited to resting pressure of the anal sphincter and RAIR) [50, 54].

Table 6.3 Awake versus under-sedation ARM [48]

	Under sedation	Awake
Resting pressure (RP)	✓	✓
Squeeze pressure (SP)		✓
Enhanced squeeze (ES)		✓
Cough reflex (CR)		✓
Push		✓
RAIR	✓	✓
Rectal sensation		✓

6.3.5 Protocol and Equipment

There are two types of manometry systems, water-perfused and solid-state either conventional or high-resolution ARM (HR-ARM). In water-perfused, the catheter is formed of multiple lumens that open at different parts of the catheter along its length according to catheter design. Water is perfused at a constant flow via a pneumohydraulic pump. External transducers detect the pressure generated by resistance to flow from lumen occlusion. In a solid-state catheter, numerous microtransducers are built into the catheter, so that pressure changes directly influence the transducers to generate electrical signal output. Either can be used for this procedure. In conventional ARM, catheters usually have fewer sensors, about three to six unidirectional sensors with wider intervals between the sensors. As such, a pull-through technique is recommended to allow accurate location of anal sphincter; this will add extra time to the total duration of the procedure compared with stationary technique used in high resolution. The output of conventional manometry catheters is in line plot. The catheters are durable, robust and not that expensive. High-resolution catheters on the other hand have several densely positioned sensors circumferentially across a defined length of the catheter. Up to 36 sensors were manufactured, and they can output either topographic colour contour or line plot. The catheters are fragile and expensive but are usually used as stationary examination, hence less time-consuming. High-resolution manometry system is significantly more expensive compared with conventional manometry. Although HRARM is described as superior to conventional ARM in adults, there is limited paediatric experience comparing the two systems.

Based on the recent BSPGHAN consensus [48] a protocol (Fig. 6.1) provides a paediatric step-by-step guide on how ARM is done in children awake. A detailed instruction guide is summarised below [48]:

- Children are instructed to defaecate if required prior to investigation.
- Children are given privacy to get ready and are asked to cover up to maintain dignity.
- Once the child is ready in the left lateral position with the knees and hips flexed, the members of the team enter the investigation room. At this stage, if play therapist input is needed, the clinician would give them the required time to prepare with the child.
- Prior to catheter insertion, perinatal inspection is carried out together with a digital rectal examination (if possible). These actions are

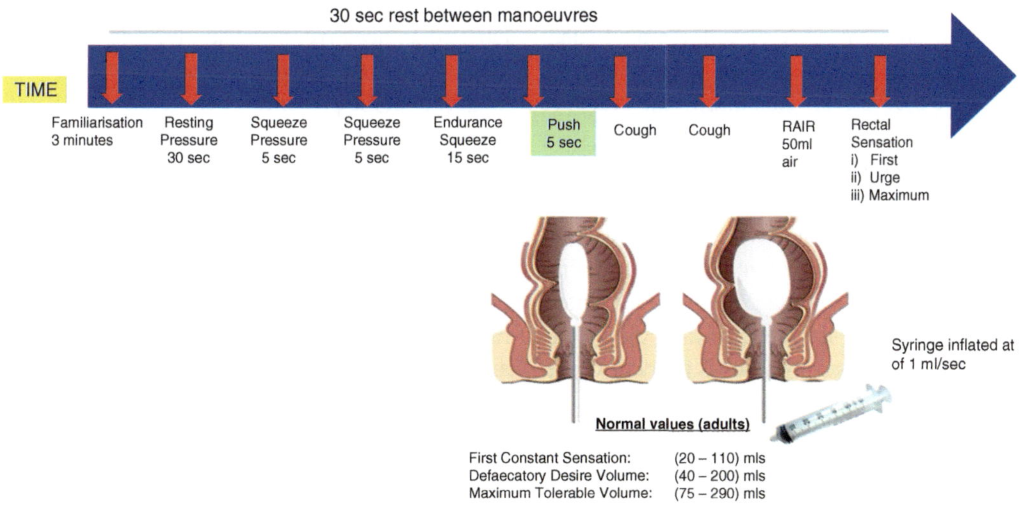

Fig. 6.1 Paediatric awake high-resolution anorectal manometry protocol (modified from adult practice for paediatric use) [44] (with permission)

desirable to assess (i) if the anal canal is filled with faeces—thus an enema or disimpaction would need to be given; (ii) general anatomy of the patient; (iii) skin excoriation; (iv) ability of the subject to understand the commands 'squeeze' and 'push'.

– The catheter is zeroed at the anal verge to calibrate and then lubricated with a manufacturer-recommended lubricant. In HRAM, the catheter is inserted into the rectum and pulled back slightly until the anal canal is located. Once the anal canal is located, the probe is taped to the buttock cheeks to avoid movement. If conventional ARM catheter is used, a pull-through is performed and several pull-through is performed to measure anal canal length.

Refer to Fig. 6.1 which describes the recent accepted UK use of paediatric ARM.

6.3.6 Traditional ARM Versus High-Resolution Anorectal Manometry (HR-ARM).

Based on the recent BSPGHAN consensus, the outcome of using ARM is a quantitative assessment of different parameters as discussed previously (Table 6.1) [44, 48, 169–171] and a qualitative assessment of the other parameters [40]. For each manoeuvre period, the anal canal area should be highlighted as an 'area of interest' using the e-sleeve box. This allows the software to derive the maximum pressure recorded over this anal length at each point in time. Averages can then be calculated automatically over the duration of the manoeuvre. The variables recorded together with their respective definitions are shown in Table 6.2. Yet, in order to interpret the results to inform clinical practice an understanding of normal values is needed, which will be discussed later in this section.

Traditional ARM in an awake child has been largely used in isolated research cases (e.g. HSCR) or specific conditions, with limited use of the novel HR-ARM [52, 172–178]. HR-ARM has enabled a paradigm shift in physiological testing of the upper gastrointestinal tract [179], replacing traditional manometry as the gold standard investigation of oesophageal function (Fig. 6.2). In adult care, HR-ARM has demonstrated its value as a dynamic investigation to visualise the anorectum as a dynamic structure during test manoeuvres (such as defaecation dynamics) [180], which leads to better appreciation of the function of the anorectum. HR-ARM is now a recognised established standard in adults [45]. There is growing evidence that, compared to the 'conventional' or 'traditional' technique, HR-ARM provides more accurate assessment of sphincter function [181] and, importantly, recto-anal coordination [182].

It is evident that HR-ARM would be able to show in detail the various subgroups of patients with dyssynergic defaecation and detect the defects of the anal sphincter at rest and during squeeze in great detail [180, 183] (Fig. 6.2). The incoordination between the relaxation of the anal sphincters and pelvic floor muscles during defaecation, called dyssynergic defaecation, may be an underlying cause of DD. This can be evaluated in real time with HR-ARM during the bear-down manoeuvre. Regardless of the type of ARM phenotype, dyssynergic defaecation leads to outlet-obstruction constipation. This diagnosis can be confirmed by the inability of expelling the balloon. In children, both the bear-down manoeuvre and the balloon expulsion test may be falsely labelled as negative, because of the lateral position adopted during the test and the anxiety to defaecate in the presence of clinicians [184]. Yet, despite the benefits of using HR-ARM, its uptake has been minimal. This can be due to several reasons such as being significantly more expensive, lack of normative data (as we see in paediatrics), limited training for clinicians who have not performed or interpreted HR-ARM within paediatrics or simply being unsure of its actual benefits and clinical implementation.

One of the principal challenges to adopting HR-ARM or conventional ARM in children is to establish new normative data sets of an adequate size for recognised measures of anal sphincter function, and to promote standardisation of the technique so that results are transferrable between

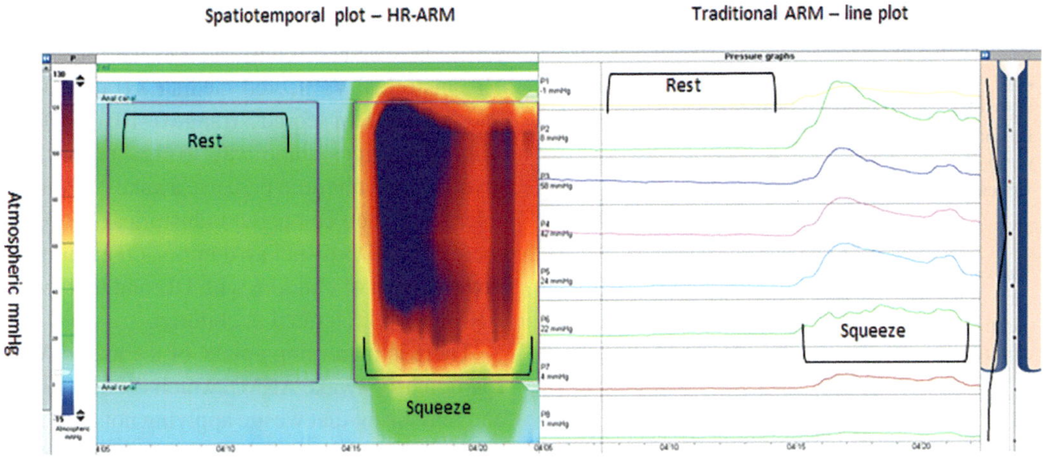

Fig. 6.2 Variation of traditional versus HR-ARM [55] (with permission)

institutions, a problem that has bedevilled traditional practice. At present, there is growing published ARM data for healthy adult controls for HR-ARM [44, 183, 185–187]. Normative data for ARM in children has primarily been produced using traditional or conventional ARM as indicated in our results, with sparse use in HR-ARM [163, 175, 188]. It is vital to establish scientific normative data sets for children, including adequate age/sex-matched size, whether under GA/sedation or awake and various physiological parameters (i.e. anal sphincter function, rectal sensation, defaecation dynamics), in order to promote standardisation of ARM within paediatrics. This will allow results to be transferrable and comparable between institutions.

At present current practice of ARM within paediatrics is making it arduous to progress our knowledge about the pathophysiological mechanism involved in defaecation disorders and furthermore advance in future treatment and research modalities in the area—just as they have in adults. Opinions have been shared regarding the need for a detailed consensus for ARM use within paediatrics [51, 169]. The recent BSPGHAN consensus has published the normative data in children that has been recommended for us according to age as guidance [48]. It was suggested that in the absence of true paediatric normal values, this working group suggest to use adult values for children over the age of 12 years and to adopt the values published by Banasiuk et al. (2016) [175] as it represents the largest paediatric series of children without lower GI symptoms. Yet the different equipment used by this group were acknowledged, as 3D HRAM may not exactly match the results from standard HRAM or conventional ARM, but the study is the largest to date in children.

6.4 Conclusions

This chapter has discussed the vast variation and lack of standardisation evident for ARM use within paediatrics. The BSPGHAN consensus has appreciated this and the need to strive for standardisation and use of consensus guidelines in the UK [48]. It is anticipated that this will further advance our understanding of the pathophysiological mechanisms involved in paediatric DD and further develop physiological phenotypes to guide and improve management.

We are still at its infancy of ARM in paediatrics, with obvious limitations (Table 6.4) in the published literature, clinical application and scientific appreciation of all the pathophysiological mechanisms that can be involved. Primarily in the UK, ARM testing in children has historically been performed under sedation [49, 50], because of the child's age, due to the likelihood that they will be uncooperative or simply because

Table 6.4 Unmet needs and gaps in knowledge [48]

Unmet needs and gaps in knowledge	Suggested strategies/actions to address this
Paediatric normative values	To study anorectal manometry parameters and assessment protocols in normal children (underway in at least one centre in the UK)
The values of each of the test parameters on children	Rectal manometry in its current format is designed to meet the clinical assessment of adult patients. Detailed studies to understand the usefulness of each of the manometry parameters in paediatrics are an urgent priority
ARM/HRAM	Comparative studies between the two systems. A multicentre study with standard protocol to minimise inter/intra-observer variabilities
Water-perfused/ solid-state catheters	Comparative studies to evaluate the practical value and utility of each of the catheter types for clinical assessment
Bowel preparation/ enema/no preparation	Comparative studies to address the need for and effect of pre-procedural preparation
Standard analysis and reporting tools	A multicentre study to evaluate inter/intra-observer variabilities and propose standardised protocol A follow-on study must be undertaken to ensure that objectives are met
Automated/ manual assessment	A multicentre study to evaluate inter/intra-observer variabilities and propose a standardised protocol A follow-on study must be undertaken to ensure that objectives are met

ARM has been perceived as an invasive procedure [51]. A recent research letter (publication pending) demonstrated in their service evaluation audit that there was a 98% success rate in performing ARM awake) [189]. Perception of discomfort/pain was recorded for awake ARM and was compared to a historical venepuncture (VP). Comparison with historical VP has several benefits: simplicity, most children had experience of VP, avoiding further discomfort, cost-effectiveness and ethical considerations. Children undergoing awake ARM were asked to compare the discomfort of awake ARM (using the validated Wong Baker scale: 0 to 10 (0—no discomfort, 10 worst discomfort/pain)) to historical VP. Historical VP scored median 7 (range 0–10) and awake ARM scored 3 (range 0–10) ($p < 0.01$). This result goes some way to reassuring those involved that children perceive awake ARM to be tolerable, when compared to historical VP. Thus, there is hope to dispel assumptions regarding the investigation of children with DD and that awake ARM is well tolerated in children.

There is increasing uptake of the use of ARM (particularly awake ARM), which is encouraging and reassuring that we are applying not only what we have learnt from the adult sector but also the significance of understanding this complex condition. This chapter has provided a current overview of ARM within paediatrics. It has highlighted the clinical benefits, but also the limitations. There is no doubt that ARM does provide advanced perception of the anorectum with insight into the pathophysiological mechanisms possibly involved in defaecation disorders in children. Nevertheless, no single investigation is able to fully characterise and comprehend all components that cause the debilitating symptoms of CC and FI [190].

Despite progression over the years, there is still a considerable way to go to gain the clinical diffusion of evolving phenotypes. Yet, it remains promising as there is evidently an abundance of interest and a matter of necessity to understand colorectal conditions within paediatrics.

References

1. Rajindrajith S, Devanarayana NM, Perera BJC, Benninga MA. Childhood constipation as an emerging public health problem. World J Gastroenterol. 2016;22(30):6864.
2. Van Den Berg MM, Benninga MA, Di Lorenzo C. Epidemiology of childhood constipation: a systematic review. Am J Gastroenterol. 2006;101(10):2401–9.
3. Mugie SM, Benninga MA, Di Lorenzo C. Epidemiology of constipation in children and adults: a systematic review Head of Department Paediatric Gastroenterology and Nutrition. Best Pract Res Clin Gastroenterol. 2011;25:3–18.
4. Boronat AC, Ferreira-Maia AP, Matijasevich A, Wang YP. Epidemiology of functional gastrointesti-

nal disorders in children and adolescents: a systematic review. World J Gastroenterol. 2017;23(21):3915.

5. Koppen IJN, Vriesman MH, Saps M, et al. Prevalence of functional defecation disorders in children: a systematic review and meta-analysis. J Pediatr. 2018;198:121–30.
6. Peeters B, Vriesman MN, Koppen IJ, Van Dijk M, Grootenhuis MA, Di Lorezo C, Benninga MA. Parental characteristics and functional constipation in children: a cross-sectional cohort study. BMJ Paediatr Open. 2017;1
7. Benninga MA, Voskuijl WP, Akkerhuis GW, Taminiau JA, Büller HA. Colonic transit times and behaviour profiles in children with defecation disorders. Arch Dis Child. 2004;89(1):13–6.
8. Athanasakos E, Ward H, Williams N, Scott S. Importance of extrasphincteric mechanisms in the pathophysiology of faecal incontinence in adults with a history of anorectal anomaly. Brit J Surg. 2008;95:1394–400.
9. Hyams JS, Di Lorenzo C, Saps M, Shulman RJ, Staiano A, Van Tilburg M. Childhood functional disorders: children and adolescents. Gastroenterology. 2016;15:S0016-5085.
10. Benninga MA, Voskuijl WP, Taminiau J. Childhood constipation: is there new light in the tunnel? J Pediatr Gastroenterol Nutr. 2004;39:448–64.
11. Koppen IJ, Nurko S, Saps M, Di Lorenzo C, Benninga MA. The pediatric Rome IV: what's new? Exp Rev Gastroenterol Hepatol. 2017;11(3):193–201.
12. Benninga MA, Büller HA, Heymans HS, Tytgat GN, Taminiau JA. Is encopresis always the result of constipation? Arch Dis Childhood. 1994;71:186–93.
13. Sommers T, Corban C, Sengupta N, et al. Emergency department burden of constipation in the United States from 2006 to 2011. Am J Gastroenterol. 2015;110:572–9.
14. NICE Guidelines; 2010. https://www.nice.org.uk/guidance
15. Voskuijl WP, Heijmans J, Heijmans HSA, Taminiau JAJM, Benninga MA. Use of Rome II criteria in childhood defecation disorders: applicability in clinical and research practice. J Pediatr. 2004;145:213–7.
16. Voskuijl WP, van der Zaag-Loonen HJ, Ketel IJ, et al. Health related quality of life in disorders of defecation: the defecation disorder list. Arch Dis Child. 2004;89:1124–7.
17. Bongers MEJ, Tabbers MM, Benninga MA. Functional nonretentive fecal incontinence in children. J Pediatr Gastroenterol Nutr. 2007;44:5–13.
18. Van Dijk M, Benninga MA, Grootenhuis MA, et al. Prevalence and associated clinical characteristics of behaviour problems in constipated children. Pediatrics. 2010;125(2):e309–17.
19. Trajanovska M, King SK, Gibb S, Goldeld S. Children who soil: a review of the assessment and management of faecal incontinence. J Paediatr Child Health. 54:1136–41.
20. Bongers ME, Benninga MA. Functional fecal incontinence in children. Annales Nestlé (English ed.). 2017;65:81–8.
21. Chumpitazi BP, Nurko S. Defecation disorders in children after surgery for Hirschsprung disease. J Pediatr Gastroenterol Nutr. 2011;53(1):75–9.
22. Nurko S, Garcia-Aranda JA, Guerrero VY, Worona LB. Treatment of intractable constipation in children: experience with cisapride. J Pediatr Gastroenterol Nutr. 1996;22:38–44.
23. Verhoef M, Lurvink M, Barf HA, et al. High prevalence of incontinence among young adults with spina bifida: description, prediction and problem perception. Spinal Cord. 2005;43:331–40.
24. Rao SSC, Benninga MA, Bharucha AE, Chiarioni G, Di Lorezo C, Whitehead WE. ANMS-ESNM position paper and consensus guidelines on biofeedback therapy for anorectal disorders. Neurogastroenterol Motil. 2015;27:594–609.
25. Wald A, Sigurdsson L. Quality of life in children and adults with constipation. Best Pract Res Clin Gastroenterol. 2011;25(1):19–27.
26. Youssef NN, Langseder AL, Verga BJ, Mones RL, Rosh JR. Chronic childhood constipation is associated with impaired quality of life: a case-controlled study. J Pediatr Gastroenterol Nutr. 2005;41(1):56–60.
27. Wang C, Shang L, Zhang Y, et al. Impact of functional constipation on health-related quality of life in preschool children and their families in Xi'an, China. PLoS One. 2013;8(10):e77273.
28. Tabbers MM, Dilorenzo C, Berger MY, et al. Evaluation and treatment of functional constipation in infants and children: evidence-based recommendations from ESPGHAN and NASPGHAN. J Pediatr Gastroenterol Nutr. 2014;58:258–74.
29. Health NCC for W and C. Constipation in children and young people: constipation child young people diagnosis manag idiopathic child constipation prim Second care; 2010.
30. Rao SCS, Meduri K. What is necessary to diagnose constipation? Best Prac Res Clin Gastroenterol. 2011;25(1):127–40.
31. Tantiphlachiva K, Rao P, Attaluri A, Rao SS. Digital rectal examination is a useful tool for identifying patients with dyssynergia. Clin Gastroenterol Hepatol. 2010;8(11):955–60.
32. Lam TJ, Felt-Bersma RJ. Clinical examination remains more important than anorectal function tests to identify treatable conditions in women with constipation. Int Urogynecol J. 2013;24(1):67–72.
33. Camilleri M, Talley NJ. Pathophysiology as a basis for understanding symptom complexes and therapeutic targets. Neurogastroenterol Motil. 2004;16(2):135–42.
34. Scott SM, van den Berg MM, Benninga MA. Rectal sensorimotor of dysfunction in constipation. Best Pract Res Clin Gastroenterol. 2011;25:103–18.
35. Gladman MA, Aziz Q, Scott SM, Williams NS, Lunniss PJ. Rectal hyposensitivity: pathophysiological mechanisms. Neurogastroenterol Motil. 2009;21:508–e5.
36. Neri L, Iovino P. Laxative Inadequate Relief Survey (LIRS) Group. Bloating is associated with worse

quality of life, treatment satisfaction, and treatment responsiveness among patients with constipation-predominant irritable bowel syndrome and functional constipation. Neurogastroenterol Motil. 2016;28(4):581–91.
37. Gowers WR. The autonomic action of the sphincter ani. Proc R Soc Med (Lond). 1877;26:77–84.
38. Azpiroz F, Enck P, Whitehead WE. Anorectal functional testing: review of collective experience. Am J Gastroenerol. 2002;97:232–40.
39. Rao SS, Patel RS. How useful are manometric tests of anorectal function in the management of defecation disorders? Am J Gastroenterol. 1997;92(3):469–75.
40. Rao SS, Azpiroz F, Diamant N, Enck P, Tougas G, Wald A. Minimum standards of anorectal manometry. Neurogastroenterol Motil. 2002;14(5):553–9.
41. Diamant NE, Kamm MA, Wald A, Whitehead WE. AGA technical review on anorectal testing techniques. Gastroenterology. 1999;116(3):735–60.
42. Potts WJ. The surgeon and the child. Philadelphia: WB Saunders Company; 1959.
43. Rao SS, Mudipalli RS, Stessman M, Zimmerman B. Investigation of the utility of colorectal function tests and Rome II criteria in dyssynergic defecation (Anismus). Neurogastroenterol Motil. 2004;16:589–96.
44. Carrington EV, Brokjær A, Craven H, et al. Traditional measures of normal anal sphincter function using high-resolution anorectal manometry (HR-ARM) in 115 healthy volunteers. Neurogastroenterol Motil. 2014;26(5):625–35.
45. Carrington EV, Heinrich H, Knowles CH, Fox M, Rao S, Altomare DF, et al. The international anorectal physiology working group (IAPWG) recommendations: Standardized testing protocol and the London classification for disorders of anorectal function. Neurogastroenterol Motil. 2020;32(1):e13679.
46. Carrington EV, Heinrich H, Knowles CH, Rao SS, Fox M, Scott SM. Methods of anorectal manometry vary widely in clinical practice: results from an international survey. Neurogastroenterol Motil. 2017;29(8):e13016.
47. Dinning PG, Carrington EV, Scott SM. The use of colonic and anorectal high-resolution manometry and its place in clinical work and in research. Neurogastroenterol Motil. 2015;27(12):1693–708.
48. Athanasakos E, Cleeve S, Thapar N, Lindley K, Perring S, Cronin H, Borrelli O, Mutalib M. Anorectal manometry in children with defecation disorders BSPGHAN Motility Working Group consensus statement. Neurogastroenterol Motil. 2020;32:e13797.
49. Keshtgar AS, Athanasakos E, Clayden GS, Ward HC. Evaluation of outcome of anorectal anomaly in childhood: the role of anorectal manometry and endosonography. Pediatr Surg Int. 2008;24(8):885–92.
50. Keshtgar AS, Choudhry MS, Kufeji D, Ward HC, Clayden GS. Anorectal manometry with and without ketamine for evaluation of defecation disorders in children. J Pediatr Surg. 2015;50(3):438–43.
51. Rodriguez L, Sood M, Di Lorenzo C, Saps M. An ANMS-NASPGHAN consensus document on anorectal and colonic manometry in children. Neurogastroenterol Motil. 2017;29(1):e12944.
52. Banasiuk M, Banaszkiewicz A, Piotrowski D, et al. 3D high-definition manometry in evaluation of children after surgery for Hirschsprung's disease: a pilot study. Adv Med Sci. 2016;61:18–22.
53. Ambartsumyan L, Rodriguez L, Morera C, Nurko S. Longitudinal and radial characteristics of intra-anal pressures in children using 3d high-definition anorectal manometry: new observations. Am J Gastroenterol. 2013;108(12):1918.
54. Pfefferkorn MD, Croffie JM, Corkins MR, et al. Impact of sedation and anesthesia on the rectoanal inhibitory reflex in children. J Pediatr Gastroenterol Nutr. 2004;38:324–7.
55. Yates G, Friedmacher F, Cleeve S, Athanasakos. Anorectal manometry in pediatric settings: a systematic review of 227 studies. Neurogastroenterol Motil. 2020;33:e14006; ahead of print.
56. Bharucha AE. Fecal incontinence. Gastroenterology. 2003;124(6):1672–85.
57. Whitehead WE, Borrud L, Goode PS, Meikle S, Mueller ER, Tuteja A, et al. Fecal incontinence in US adults: epidemiology and risk factors. Gastroenterology. 2009;137(2):512–7.
58. Rao SS. Advances in diagnostic assessment of fecal incontinence and dyssynergic defecation. Clin Gastroenterol Hepatol. 2010;8(11):910–9.
59. Salvioli B, Bharucha AE, Rath-Harvey D, Pemberton JH, Phillips SF. Rectal compliance, capacity, and rectoanal sensation in fecal incontinence. Am J Gastroenterol. 2001;96:2158–68.
60. Read NW, Bartolo DC, Read MG. Differences in anal function in patients with incontinence to solids and in patients with incontinence to liquids. Br J Surg. 1984;71:39–42.
61. Engel AF, Kamm MA, Bartram CI, Nicholls RJ. Relationship of symptoms in faecal incontinence to specific sphincter abnormalities. Int J Colorectal Dis. 1995;10:152–5.
62. Vaizey CJ, Kamm MA, Bartram CI. Primary degeneration of the internal anal sphincter as a cause of passive faecal incontinence. Lancet. 1997;349:612–5.
63. Deutekom M, Dobben AC, Terra MP, Engel AF, Stoker J, Bossuyt PM, Boeckxstaens GE. Clinical presentation of fecal incontinence and anorectal function: what is the relationship? Am J Gastroenterol. 2007;102:351–61.
64. Gee AS, Durdey P. Urge incontinence of faeces is a marker of severe external anal sphincter dysfunction. Br J Surg. 1995;82:1179–82.
65. Chan CL, Lunniss PJ, Wang D, Williams NS, Scott SM. Rectal sensorimotor dysfunction in patients with urge faecal incontinence: evidence from prolonged manometric studies. Gut. 2005;54:1263–72.
66. Vasudevan SP, Scott SM, Gladman MA, Lunniss PJ. Rectal hyposensitivity: evaluation of anal sensation in female patients with refractory con-

stipation with and without faecal incontinence. Neurogastroenterol Motil. 2007;19:660–7.
67. Bharacha AE. Pelvic floor: anatomy and function. Neurogastroenterol and Motility. 2006;18:507–19.
68. Burgell R, Scott M. Rectal hyposensitivity. J Neurogastroenterol Motil. 2012;18:373–84.
69. Gladman MA, Scott SM, Chan CL, Williams NS, Lunniss PJ. Rectal hyposensitivity: prevalence and clinical impact in patients with intractable constipation and fecal incontinence. Dis Colon Rectum, 2003; 46(2): 238-246.
70. Keshtgar AS, Ward HC, Clayden GS. Pathophysiology of chronic childhood constipation: functional and morphological evaluation by anorectal manometry and endosonography and colonic transit study. J Pediatr Surg. 2013;48:806–12.
71. Callaghan RP, Nixon HH. Megarectum: physiological observations. Arch Dis Child. 1964;39:153–7.
72. Gladman MA, Dvorkin LS, Lunniss PJ, Williams NS, Scott SM. Rectal hyposensitivity: a disorder of the rectal wall or the afferent pathway? An assessment using the barostat. Am J Gastroenterol. 2005;100:106–14.
73. Van der Plas RN, Benninga MA, Staalman CR, Akkermans LM, Redekop WK, Taminiau JA, Büller HA. Megarectum in constipation. Arch Dis Childhood. 2000;83(1):52–8.
74. Fathy A, Megahed A, Barakat T, Abdalla AF. Anorectal functional abnormalities in Egyptian children with chronic functional constipation. Arab J Gastroenterol. 2013;14(1):6–9.
75. Yuan Z, Cheng W, Hou A, Wang W, Zhang S, Liu D, Gao F, Li H, Wang W. Constipation is associated with spina bifida occulta in children. Clin Gastroenterol Hepatol. 2008;6(12):1348–53.
76. Pakarinen MP, Koivusalo A, Rintala RJ. Functional fecal soiling without constipation, organic cause or neuropsychiatric disorders? J Pediatr Gastroenterol Nutr. 2006;43(2):206–8.
77. van Ginkel R, Benninga MA, Blommaart PJ, van der Plas RN, Boeckxstaens GE, Büller HA, Taminiau JA. Lack of benefit of laxatives as adjunctive therapy for functional nonretentive fecal soiling in children. J Pediatr. 2000;137(6):808–13.
78. Molnar D, Taitz LS, Urwin OM, Wales JK. Anorectal manometry results in defecation disorders. Arch Dis Childhood. 1983;58(4):257–61.
79. Loening-Baucke VA. Abnormal rectoanal function in children recovered from chronic constipation and encopresis. Gastroenterology. 1984;87:1299–304.
80. Meunier P, Marechal JM, de Beaujeu MJ. Rectoanal pressures and rectal sensitivity studies in chronic childhood constipation. Gastroenterology. 1979;77:330–6.
81. Clayden GS, Lawson JO. Investigation and management of longstanding chronic constipation in childhood. Arch Dis Child. 1976;51:918–23.
82. Keshtgar AS, Ward HC, Clayden GS, et al. Thickening of the internal anal sphincter in idiopathic constipation in children. Pediatr Surg Int. 2004;20:817–23.
83. Loening-Baucke VA, Younoszai MK. Abnormal and sphincter response in chronically constipated children. J Pediatr. 1982;100:213–8.
84. Van der Plas RN, Benninga MA, Redekop WK, Taminiau JA, Büller HA. Randomised trial of biofeedback training for encopresis. Arch Dis Childhood. 1996;75:367–74.
85. Cucchiara S, Coremans G, Staiano A, et al. Gastrointestinal transit time and anorectal manometry in children with fecal soiling. J Pediatr Gastroenterol Nutr. 1984;3:545–50.
86. Van Ginkel R, Büller HA, Boeckxstaens GE, et al. The effect of anorectal manometry on the outcome of treatment in severe childhood constipation: a randomized, controlled trial. Pediatrics. 2001;108(1):E9.
87. Loening-Baucke V. Biofeedback treatment for chronic constipation and encopresis in childhood: long-term outcome. Pediatrics. 1995;96:105–10.
88. Arhan P, Devroede G, Jehannin B, et al. Idiopathic disorders of fecal continence in children. Pediatrics. 1983;71(5):774–9.
89. Rao SS. Dyssynergic defecation and biofeedback therapy. Gastroenterol Clin N Am. 2008;37:569–86.
90. Harraf F, Schmulson M, Saba L, Niazi N, Fass R, Munakata J, et al. Subtypes of constipation predominant irritable bowel syndrome based on rectal perception. Gut. 1998;43:388–94.
91. Mayer EA, Naliboff BD, Craig AD. Neuroimaging of the brain-gut axis: from basic understanding to treatment of functional GI disorders. Gastroenterology. 2006;131:1925–42.
92. Bongers ME, van Wijk MP, Reitsma JB, et al. Long-term prognosis for childhood constipation: clinical outcomes in adulthood. Pediatrics. 2010;126:e156–62.
93. Chitkara DK, Bredenoord AJ, Cremonini F, et al. The role of pelvic floor dysfunction and slow transit constipation in adolescents with chronic constipation. Am J Gastroenterol. 2004;99:1579–84.
94. Borowitz SM, Sutphen J, Ling W, Cox DJ. Lack of correlation of anorectal manometry with symptoms of chronic childhood constipation and encopresis. Dis Colon Rectum. 1996;39:400–5.
95. Feinberg L, Mahajan L, Steffen R. The constipated child: is there a correlation between symptoms and manometric findings? J Pediatr Gastroenterol Nutr. 2008;47:607–11.
96. Clayden G, Agnarsson U. Constipation in childhood. Oxford Med Publ. 1991;
97. Adıgüzel Ü, Ağengin K, Kırıştıoğlu I, Doğruyol H. Transanal endorectal pull-through for Hirschsprung's disease: experience with 50 patients. Irish J Med Sci. 2017;186:433–7.
98. Tang PM, Leung MW, Hung JW, Chung KL, Liu CS, Chao NS, Liu KK. Application of three-dimensional high resolution anorectal manometry to demonstrate anal sphincter pressure asymmetry in children after anorectal surgery: a pilot study. HK J Paediatr (New Series). 2017;22:204–8.

99. Chung PH, Wong KK, Leung JL, Tam PK, Chung KL, Leung MW, Chao NS, Liu KK, Tsui BS, Chan EK, Tam PY. Clinical and manometric evaluations of anorectal function in patients after transanal endorectal pull-through operation for Hirschsprung's disease: a multicentre study. Surg Pract. 2015;19(3):113–9.
100. Zhang JS, Li L, Hou WY, Liu SL, Diao M, Zhang J, Ming AX, Cheng W. Transanal rectal mucosectomy and partial internal anal sphincterectomy for Hirschsprung's disease. J Pediatr Surg. 2014;49:831–4.
101. Wu XJ, Zhang HY, Li N, Yan MS, Wei J, Yu DH, Feng JX. A new diagnostic scoring system to differentiate Hirschsprung's disease from Hirschsprung's disease-allied disorders in patients with suspected intestinal dysganglionosis. Int J Colorectal Dis. 2013;28:689–96.
102. Demirbag S, Tiryaki T, Purtuloglu T. Importance of anorectal manometry after definitive surgery for Hirschsprung's disease in children. Afr J Paediatr Surg. 2013;10(1):1.
103. Zakaria OM. Bowel function and fecal continence after Soave's trans-anal endorectal pull-through for Hirschsprung's disease: a local experience. Updates Surg. 2012;64:113–8.
104. Sun X, Wang R, Zhang L, Li D, Li Y. Efficacy of pelvic floor muscle training for the treatment of fecal incontinence after Soave procedure for Hirschsprung disease. Eur J Pediatr Surg. 2012;22:300–4.
105. Martinac M, Ebling B, Dujšin M, Majstorović K, Jurčić D, Bilić A, Troskot PR. Clinical and anorectal motility features in chronically constipated children. Collegium Antropologicum. 2011;35:505–12.
106. El-Hak NA, El-Hemaly MM, Negm EH, El-Hanafy EA, Messeh MH, Bary HH. Functional outcome after Swenson's operation for Hirschsprung's disease. Saudi J Gastroenterol. 2010;16:30.
107. Zhang HY, Feng JX, Huang L, Wang G, Wei MF, Weng YZ. Diagnosis and surgical treatment of isolated hypoganglionosis. World J Pediatr. 2008;4:295–300.
108. Huang Y, Zheng S, Xiao X. A follow-up study on postoperative function after a transanal Soave 1-stage endorectal pull-through procedure for Hirschsprung's disease. J Pediatr Surg. 2008;43(9):1691–5.
109. Lee JH, Choe YH, Lee SK, Seo JM, Kim JH, Suh YL. Allergic proctitis and abdominal distention mimicking Hirschsprung's disease in infants. Acta Paediatr. 2007;96(12):1784–9.
110. Yoshino H, Kayaba H, Hebiguchi T, Morii M, Hebiguchi T, Itoh W, Chihara J, Kato T. Anal ultraslow waves and high anal pressure in childhood: a clinical condition mimicking Hirschsprung disease. J Pediatr Surg. 2007;42(8):1422–8.
111. Zhang SC, Bai YZ, Wang W, Wang WL. Long-term outcome, colonic motility, and sphincter performance after Swenson's procedure for Hirschsprung's disease: a single-center 2-decade experience with 346 cases. Am J Surg. 2007;194:40–7.
112. Messina M, Amato G, Meucci D, Molinaro F, Nardi N. Topical application of isosorbide dinitrate in patients with persistent constipation after pull-through surgery for Hirschsprung's disease. Eur J Pediatr Surg. 2007;17:62–5.
113. Tiryaki T, Demirbag S, Atayurt H, Çetinkursun S. Topical nitric oxide treatment after pull through operations for Hirschsprung disease. J Pediatr Gastroenterol Nutr. 2005;40:390–2.
114. Peterlini FL, Martins JL. Modified transanal rectosigmoidectomy for Hirschsprung's disease: clinical and manometric results in the initial 20 cases. J Pediatr Surg. 2003;38:1048–50.
115. Keshtgar A, Ward H, Clayden G, de Sousa N. Investigations for incontinence and constipation after surgery for Hirschsprung's disease in children. Pediatr Surg Int. 2003;19(1-2):4–8.
116. Millar AJ, Steinberg RM, Raad J, Rode H. Anal achalasia after pull-through operations for Hirschsprung's disease-preliminary experience with topical nitric oxide. Eur J Pediatr Surg. 2002;12:207–11.
117. Miele E, Tozzi A, Staiano A, Toraldo C, Esposito C, Clouse RE. Persistence of abnormal gastrointestinal motility after operation for Hirschsprung's disease. Am J Gastroenterol. 2000;95:1226–30.
118. Niedzielski J. Posterior sagittal abdominoperineal pull-through: a new approach to definitive treatment of Hirschsprung's disease—initial experience. J Pediatr Surg. 1999;34:572–6.
119. Banani SA, Forootan H. Role of anorectal myectomy after failed endorectal pull-through in Hirschsprung's disease. J Pediatr Surg. 1994;29:1307–9.
120. Moore SW, Millar AJ, Cywes S. Long-term clinical, manometric, and histological evaluation of obstructive symptoms in the postoperative Hirschsprung's patient. J Pediatr Surg. 1994;29:106–11.
121. Low PS, Quak SH, Prabhakaran K, Joseph VT, Chiang GS, Aiyathurai EJ. Accuracy of anorectal manometry in the diagnosis of Hirschsprung's disease. J Pediatr Gastroenterol Nutr. 1989;9:342–6.
122. Sumitomo K, Ikeda K, Nagasaki A. The use of prostaglandin F2α and scopolamine-N-butylbromide in anorectal manometric diagnosis. Zeitschrift für Kinderchirurgie. 1986;41:344–7.
123. Nagasaki A, Ikeda K, Hayashida Y. Radiologic diagnosis of Hirschsprung's disease utilizing rectosphincteric reflex. Pediatr Radiol. 1984;14:384–7.
124. Rosenberg AJ, Vela AR. A new simplified technique for pediatric anorectal manometry. Pediatrics. 1983;71:240–5.
125. Faverdin C, Dornic C, Arhan P, Devroede G, Jehannin B, Revillon Y, Pellerin D. Quantitative analysis of anorectal pressures in Hirschsprung's disease. Dis Colon Rectum. 1981;24:422–7.
126. Nagasaki A, Ikeda K, Suita S. Postoperative sequential anorectal manometric study of children with Hirschsprung's disease. J Pediatr Surg. 1980;15:615–9.

127. Suzuki H, Amano SI, Honzumi M, Saijo H, Sakakura K. Rectoanal pressures and rectal compliance in constipated infants and children. Zeitschrift für Kinderchirurgie. 1980;29:330–6.
128. Varma KK. Experience in the diagnosis of Hirschsprung's disease using the anomanometer. J Pediatr Surg. 1979;14(5):551–3.
129. Meunier P, Mollard P, de Beaujeu JM. Manometric studies of anorectal disorders in infancy and childhood: an investigation of the physiopathology of continence and defaecation. Br J Surg. 1976;63:402–7.
130. Meunier P, Mollard P. Control of the internal anal sphincter (manometric study with human subjects). Pflügers Arch. 1977;370:233–9.
131. Heikenen JB, Werlin SL, Di Lorenzo C, Hyman PE, Cocjin J, Flores AF, et al. Colonic motility in children with repaired imperforate anus. Dig Dis Sci. 1999;44:1288–92.
132. Rintala R, Lindahl H, Marttinen E, Sariola H. Constipation is a major functional complication after internal sphincter-saving posterior sagittal anorectoplasty for high and intermediate anorectal malformations. J Pediatr Surg. 1993;28:1054–8.
133. Rintala RJ, Lindahl H. Is normal bowel function possible after repair of intermediate and high anorectal malformations? J Pediatr Surg. 1995;30:491–4.
134. Ong NT, Beasley SW. Long-term continence in patients with high and intermediate anorectal anomalies treated by sacroperineal (Stephens) rectoplasty. J Pediatr Surg. 1991;26:44–8.
135. Kumar S, Ramadan SA, Gupta V, Helmy S, Debnath P, Alkholy A. Use of anorectal manometry for evaluation of postoperative results of patients with anorectal malformation: a study from Kuwait. J Pediatr Surg. 2010;45:1843–8.
136. Keshtgar AS, Ward HC, Sanei A, Clayden GS. Botulinum toxin, a new treatment modality for chronic idiopathic constipation in children: long-term follow-up of a double-blind randomized trial. J Pediatr Surg. 2007;42:672–80.
137. Iwai N, Iwata G, Kimura O, Yanagihara J. Is a new biofeedback therapy effective for fecal incontinence in patients who have anorectal malformations? J Pediatr Surg. 1997;32:1626–9.
138. Iwai N, Kume Y, Kimura O, Ono S, Aoi S, Tsuda T. Effects of herbal medicine Dai-Kenchu-to on anorectal function in children with severe constipation. Eur J Pediatr Surg. 2007;17:115–8.
139. Arhan P, Faverdin C, Devroede G, Besancon-Lecointe I, Fekete C, Goupil G, Black R, Jan D, Martelli H, Pellerin D. Biofeedback re-education of faecal continence in children. Int J Colorectal Dis. 1994;9:128–33.
140. Van der Plas RN, Benninga MA, Büller HA, Bossuyt PM, Akkermans LM, Redekop WK, Taminiau JA. Biofeedback training in treatment of childhood constipation: a randomised controlled study. Lancet. 1996;348:776–80.
141. Benninga MA, Büller HA, Taminiau JA. Biofeedback training in chronic constipation. Arch Dis Childhood. 1993;68:126–9.
142. Loening-Baucke V. Modulation of abnormal defecation dynamics by biofeedback treatment in chronically constipated children with encopresis. J Pediatr. 1990;116:214–22.
143. Loening-Baucke V. Persistence of chronic constipation in children after biofeedback treatment. Dig Dis Sci. 1991;36:153–60.
144. Shepherd K, Hickstein R, Rose V, Nasser C, Cleghorn GJ, Shepherd RW. Faecal incontinence in childhood: a multidisciplinary approach including biofeedback. J Paediatr Child Health. 1989;25:351–5.
145. Olmsted RW, Olness K, McParland FA, Piper J. Biofeedback: a new modality in the management of children with fecal soiling. J Pediatr. 1980;96:505–9.
146. Wald A, Chandra R, Gabel S, Chiponis D. Evaluation of biofeedback in childhood encopresis. J Pediatr Gastroenterol Nutr. 1987;6:554–8.
147. Nolan T, Catto-Smith T, Coffey C, Wells J. Randomised controlled trial of biofeedback training in persistent encopresis with anismus. Arch Dis Childhood. 1998;79:131–5.
148. Caruso AM, Catalano P, Voti GL, Salerno S, Casuccio A, Pace MRD, Cimador M. Prognostic evaluation of biofeedback response in patients treated for anorectal malformation. J Pediatr Surg. 2015;50:1648–52.
149. Voskuijl WP, van Ginkel R, Benninga MA, Hart GA, Taminiau JA, Boeckxstaens GE. New insight into rectal function in pediatric defecation disorders: disturbed rectal compliance is an essential mechanism in pediatric constipation. J Pediatr. 2006;148:62–7.
150. Kajbafzadeh AM, Sharifi-Rad L, Nejat F, Kajbafzadeh M, Talaei HR. Transcutaneous interferential electrical stimulation for management of neurogenic bowel dysfunction in children with myelomeningocele. Int J Colorectal Dis. 2012;27:453–8.
151. Siddiqui A, Rosen R, Nurko S. Anorectal manometry may identify children with spinal cord lesions. J Pediatr Gastroenterol Nutr. 2011;53:507–11.
152. Tsuda T, Iwai N, Kimura O, Kubota Y, Ono S, Sasaki Y. Bowel function after surgery for anorectal malformations in patients with tethered spinal cord. Pediatr Surg Int. 2007;23:1171–4.
153. Kayaba H, Hebiguchi T, Itoh Y, Yoshino H, Mizuno M, Morii M, Adachi T, Chihara J, Kato T. Evaluation of anorectal function in patients with tethered cord syndrome: saline enema test and fecoflowmetry. J Neurosurg Spine. 2003;98:251–7.
154. Simpser E, Kahn E, Kenigsberg K, Duffy L, Markowitz J, Daum F. Neuronal intestinal dysplasia: quantitative diagnostic criteria and clinical management. J Pediatr Gastroenterol Nutr. 1991;12:61–4.
155. Bustorff-Silva JM, Costa-Pinto EA, Fukushima E. Role of anorectal manometry in the differential diagnosis of chronic constipation in children. J Pediatr (Rio J). 2000;76:227–32.
156. Paskins JR, Lawson JO, Clayden GS. The effect of ketamine anesthesia on anorectal manometry. J Pediatr Surg. 1984;19:289–91.
157. Huang YH. Neurogenic bowel: dysfunction and rehabilitation. In: Cifu DX, Lew HL, editors. Braddom's

clinical handbook of physical medicine and rehabilitation. Amsterdam: Elsevier; 2018. p. 143–149.e7.
158. Martins EC, Peterlini FL, Fagundes DJ, Martins JL. Clinical, manometric and profilometric evaluation after surgery for Hirschsprung's disease: comparison between the modified Duhamel and the transanal rectosigmoidectomy techniques. Acta Cirurgica Brasileira. 2009;24:416–22.
159. Marques GM, Martins JL, Nobre VD. Comparison between perfusion and balloon techniques for performing anorectal manometry in children with intestinal constipation. Acta Cirurgica Brasileira. 2008;23:405–11.
160. Vital Júnior PF, Martins JL, Peterlini FL. Posterior sagittal anorectoplasty in anorectal anomalies: clinical, manometric and profilometric evaluation. Sao Paulo Med J. 2007;125:163–9.
161. Li ZH, Dong M, Wang ZF. Functional constipation in children: investigation and management of anorectal motility. World J Pediatr. 2008;4:45–8.
162. Bigélli RH, Fernandes MI, Vicente YA, Dantas RO, Galvão LC, Campos AD. Anorectal manometry in children with chronic functional constipation. Arquivos de Gastroenterologia. 2005;42:178–81.
163. Benninga MA, Wijers OB, Taminiau JA, Klopper PJ, Tytgat GN, Akkermans LM. Manometry, profilometry, and endosonography: normal physiology and anatomy of the anal canal in healthy children. J Pediatr Gastroenterol Nutr. 1994;18:68–77.
164. Rodriguez L, Siddiqui A, Nurko S. Internal anal sphincter relaxation associated with bisacodyl-induced colonic high amplitude propagating contractions in children with constipation: a coloanal reflex? Neurogastroenterol Motil. 2012;24:1023–e545.
165. Raghunath N, Glassman MS, Halata MS, Berezin SH, Stewart JM, Medow MS. Anorectal motility abnormalities in children with encopresis and chronic constipation. J Pediatr. 2011;158:293–6.
166. Frenckner B, Euler CV. Influence of pudendal block on the function of the anal sphincters. Gut. 1975;16:482–9.
167. Tran K, Kuo B, Zibaitis A, Bhattacharya S, Cote C, Belkind-Gerson J. Effect of propofol on anal sphincter pressure during anorectal manometry. J Pediatr Gastroenterol Nutr. 2014;58:495–7.
168. Liu TT, Yi CH, Chen CL, Liu HT, Chen TY. Influence of intravenous propofol sedation on anorectal manometry in healthy adults. Am J Med Sci. 2009;337:429–31.
169. Di Lorenzo C, Hillemeier C, Hyman P, et al. Manometry studies in children: minimum standards for procedures. Neurogastroenterol Motil. 2002;14:411–20.
170. Lee TH, Bharucha AE. How to perform and interpret a high-Resolution anorectal manometry test. J Neurogastroenterol Motil. 2015;22:46–59.
171. Coss-Adame E, Rao SSC, Valestin J, Ali-Azamar A, Remes-Troche JM. Accuracy and reproducibility of high-definition anorectal manometry and pressure topography analyses in healthy subjects. Clin Gastroenterol Hepatol. 2015;13:1143–50.
172. Huang Y, Shen C, Zheng S, Xiao X. A follow-up study on postoperative function after a modified semi-posterior sagittal one-stage anorectoplasty for female imperforate anus with a rectovesibula fistula. Int J Clin Exp Med. 2017;10:913–20.
173. Tang YF, Chen JG, An HJ, et al. High-resolution anorectal manometry in newborns: normative values and diagnostic utility in Hirschsprung disease. Neurogastroenterol Motil. 2014;26:1565–72.
174. Wang Z, Hu L, Jin X, Li X, Xu L. Evaluation of postoperative anal functions using endoanal ultrasonography and anorectal manometry in children with congenital anorectal malformations. J Pediatr Surg. 2016;51:416–20.
175. Banasiuk M, Banaszkiewicz A, Dziekiewicz M, Załęski A, Albrecht P. Values from three-dimensional high-resolution anorectal manometry analysis of children without lower gastrointestinal symptoms. Clin Gastroenterol Hepatol. 2016;14:993–1000.
176. Meinds RJ, Trzpis M, Broens PM. Anorectal manometry may reduce the number of rectal suction biopsy procedures needed to diagnose Hirschsprung disease. J Pediatr Gastroenterol Nutr. 2018;67:322–7.
177. Wu JF, Lu CH, Yang CH, Tsai IJ. Diagnostic role of anal sphincter relaxation integral in high-resolution anorectal manometry for Hirschsprung disease in infants. J Pediatr. 2018;194:136–41.
178. Sintusek P, Rybak A, Mutalib M, Thapar N, Borrelli O, Lindley KJ. Preservation of the colo-anal reflex in colonic transection and post-operative Hirschsprung's disease: potential extrinsic neural pathway. Neurogastroenterol Motil. 2019;31(1):e13472.
179. Bredenoord AJ, Fox M, Kahrilas PJ, Pandolfino JE, Schwizer W, Smout AJPM. Chicago classification criteria of esophageal motility disorders defined in high resolution esophageal pressure topography. Neurogastroenterol Motil. 2012;24:57–65.
180. Ratuapli SK, Bharucha AE, Noelting J, Harvey DM, Zinsmeister AR. Phenotypic identification and classification of functional defecatory disorders using high-resolution anorectal manometry. Gastroenterology. 2013;144:314–22.
181. Carrington EV, Knowles CH, Grossi U, Scott SM. High-resolution anorectal manometry measures are more accurate than conventional measures in detecting anal hypocontractility in women with fecal incontinence. Clin gastroenterol Hepatol. 2019;17:477–85.
182. Sauter M, Heinrich H, Fox M, Misselwitz B, Halama M, Schwizer W, et al. Toward more accurate measurements of anorectal motor and sensory function in routine clinical practice: validation of high-resolution anorectal manometry and Rapid Barostat Bag measurements of rectal function. Neurogastroenterol Motil. 2014;26:685–95.

183. Noelting J, Ratuapli SK, Bharucha AE, Harvey DM, Ravi K, Zinsmeister AR. Normal values for high-resolution anorectal manometry in healthy women: effects of age and significance of rectoanal gradient. Am J Gastroenterol. 2012;107:1530–6.
184. Belkind-Gerson J, Goldstein AM, Kuo B. Balloon expulsion test as a screen for outlet obstruction in children with chronic constipation. J Pediatr Gastroenterol Nutr. 2013;56:23–6.
185. Li Y, Yang X, Xu C, Zhang Y, Zhang X. Normal values and pressure morphology for three-dimensional high-resolution anorectal manometry of asymptomatic adults: a study in 110 subjects. Int J Colorectal Dis. 2013;28:1161–8.
186. Gundling F, Seidl H, Scalercio N, Schmidt T, Schepp W, Pehl C. Influence of gender and age on anorectal function: normal values from anorectal manometry in a large Caucasian population. Digestion. 2010;81:207–13.
187. Carrington EV, Grossi U, Knowles CH, Scott SM. Normal values for high-resolution anorectal manometry: a time for consensus and collaboration. Neurogastroenterol Motil. 2014;26:1356–7.
188. Kumar S, Ramadan S, Gupta V, Helmy S, Atta I, Alkholy A. Manometric tests of anorectal function in 90 healthy children: a clinical study from Kuwait. J Pediatr Surg. 2009;44:1786–90.
189. Athanasakos E and Cleeve S. An audit on awake anorectal manometry: tolerability in children. Archives Dis Child; Publication Pending (Research Letter).
190. Athanasakos E, Dalton S, McDowell S, Shea T, Blakeley K, Rawat D, Cleeve S. Scientific solution to a complex problem: physiology and multidisciplinary team improve understanding and outcome in chronic constipation and faecal incontinence. Pediatr Surg Int. 2020;36:295–303.

Colonic Manometry

7

Renato Tambucci and Osvaldo Borrelli

7.1 Introduction

The colon plays a crucial role in human homeostasis. Colonic activity is essential in regulating the frequency of defecation and consistency of stools; moreover, it serves essential functions such as absorption of water and electrolytes and salvages nutrients after bacterial metabolism of complex carbohydrates. Normal propagation of feces through the colon is critical for maintaining the physiology of defecation. Continence and defecation are inextricably linked and require a coordinated involvement from the enteric muscles, the nervous system (autonomic and somatic), and the musculoskeletal system. Colonic motility is predominantly involuntary and primarily mediated by the enteric nervous system in association with parasympathetic and sympathetic inputs [1]. Although there is still limited knowledge about the normal motor patterns, it is clear that altered colonic motility causes abnormal propulsion of the feces. This may interfere with the colon's physiological functions and result in gastrointestinal disorders such as diarrhea, irritable bowel syndrome, or constipation [2]. The measurement of colonic transit time (by radiopaque marker test or scintigraphy) provides crucial information about the nature of the patient's constipation and fecal incontinence. Indeed, it offers determinant data by differentiating between normal-transit constipation, slow-transit constipation, and outlet obstruction, and between non-retentive and overflow constipation-associated fecal incontinence. However, the colonic transit time assessment is an indirect estimation of the colonic motility since it cannot provide data on the contractile activity of the colon. Colonic manometry assesses the colon's neuromuscular integrity by measuring intraluminal pressures and coordination of the colonic muscle pressure activity [2–5].

7.2 Colonic Manometry

Colonic manometry by recording motility directly offers valuable information about the colonic motor function [6]. Although the introduction of recent innovations in colonic manometry methodology has implemented our awareness about the normal motor patterns of the colon, its real role in clinical practice is not definitely established [2–5, 7].

Nonetheless, colonic manometry assessment has been recommended by guidelines in children

R. Tambucci (✉)
Digestive Endoscopy and Surgery Unit, Bambino Gesù Children's Hospital, IRCCS, Rome, Italy
e-mail: renato.tambucci@opbg.net

O. Borrelli
Division of Neurogastroenterology & Motility, Department of Pediatric Gastroenterology, Great Ormond Street Hospital, London, UK
e-mail: osvaldo.borrelli@gosh.nhs.uk

M. Lima, G. Ruggeri (eds.), *Ano-Rectal Endosonography and Manometry in Paediatrics*,
https://doi.org/10.1007/978-3-030-97668-2_7

and adults with treatment-refractory constipation and who have evidence of slow colonic transit in the absence of an evacuatory disorder before consideration of surgical intervention [4, 5, 8, 9].

7.2.1 Methodology

The colon's motor function is assessed by colonic manometry using catheters placed within the colonic lumen equipped with pressure-recording ports placed along the catheter length. Over the last decade, manometric techniques have rapidly progressed from systems using a water-perfused catheter with few pressure-recording channels to the systems using solid-state catheters, which enable more detailed definition along with colonic segments by a substantial increase of the pressure-recording ports (high-resolution manometry, HRM) [2, 10]. Moreover, computer processing advances have greatly improved the pressure data presentation, leading to a more accurate and intuitive spatiotemporal mapping of the colon's motor function [11]. The catheter type, number of recording sites, and spacing between sensors differ amongst centers. There are currently two different main types of manometric catheters available on the market: water-perfused and solid-state [3, 7]. Water-perfused catheters require a pneumohydraulic pump to continuously perfuse distilled water through the side holes of the catheter. The probe incorporates between 4 and 20 recording side holes, with an inter-side hole distance of between 1 and 15 cm. Recording may be displayed as a color contour plot in the newest water-perfused systems that different centers are still employing. The solid-state manometry catheters consist of strain gauges embedded into a flexible PVC catheter, generally containing up to 36 recording sites closely spaced [3].

7.2.1.1 Catheter Placement Techniques

Patients are advised to discontinue medications affecting gastrointestinal motility before the test. In children, the catheter is placed into the colon only via the anus in a retrograde fashion, unlike in adults in which anterograde placement (through the nose or mouth) has been described. In the presence of ostomies (e.g., cecostomy), the catheter placement can be done via the ostomy in an antegrade manner [12, 13]. Under general anesthesia, retrograde placement can be achieved either endoscopically or using fluoroscopic guidance. After adequate bowel preparation, endoscopic placement is achieved by advancing the catheter (grasped by biopsy forceps or hemostatic clips) along with the colonoscope (Fig. 7.1). Single balloon-assisted colonoscopy for placing colonic manometry catheters has been proposed to overcome the technical difficulty that might be encountered in constipated children with sigmoid and other colonic segment redundancy [14].

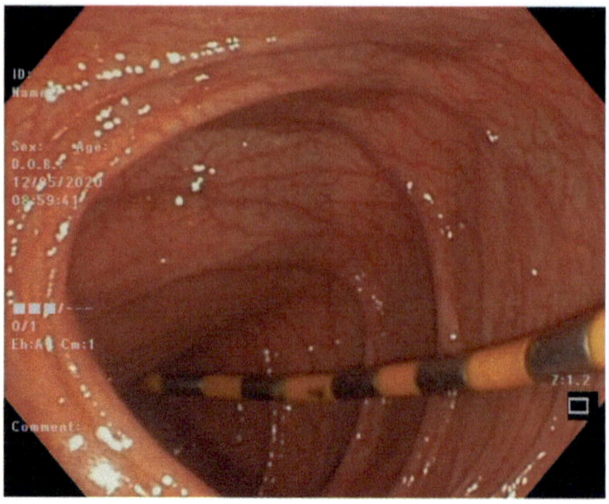

Fig. 7.1 Endoscopic image of a solid-state catheter for colonic manometry advanced through the colon

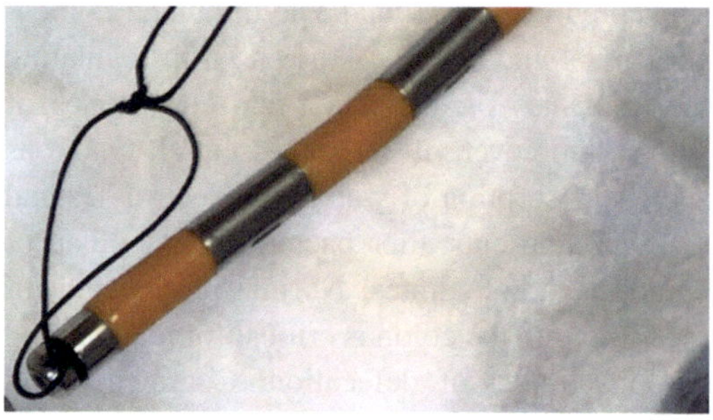

Fig. 7.2 Suture loop at the tip of the catheter

The catheter should be positioned with the tip at or, preferably, beyond the hepatic flexure; in order to minimize the risk of dislodgement during testing, the catheter can be anchored over a suture loop located at the tip (Fig. 7.2) with hemostatic endoclips deployed on a mucosal fold (Fig. 7.3) [12, 13]. After placement, the position of the catheter should be radiologically verified (Fig. 7.4).

7.2.1.2 Study Protocol

Despite variability in study protocols between centers, the manometric recording is usually started 2–4 h after the catheter placement once the patient has fully recovered from anesthesia. In clinical practice, studies last approximately 4–8 h, covering 1–2 h before and after a test meal and 1–2 h after pharmacologic stimuli. 24-Hour ambulatory monitoring while patients are encouraged to return to normal activities, including eating and sleeping, may provide more physiologic data but is mainly restricted to research purposes [12, 13].

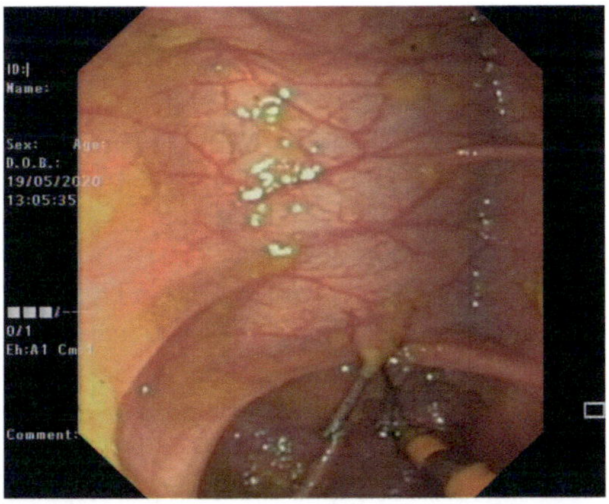

Fig. 7.3 Endoscopic image of a solid-state catheter anchored with hemostatic endoclips deployed on a mucosal fold

After the effect of the anesthesia, while the patient is comfortably lying or sitting on the bed, the study initiates by recording baseline colonic motility during 1–2 h fasting period. Therefore, postprandial motility is recorded after ingestion of a combined semisolid meal (20 kcal/kg) to evaluate the presence of the so-called gastrocolic reflex characterized by an increase in phasic and tonic motor activity. The last part of colonic manometry is a provocative test with stimulant medication administration. It is usually performed with bisacodyl (0.2–0.4 mg/kg; max 20 mg) delivered directly into the colon via the catheter's central lumen or via an ostomy opening when present. During the entire study, the patients' activities and symptoms should be carefully noted using the recording software's dedicated utilities [12, 13].

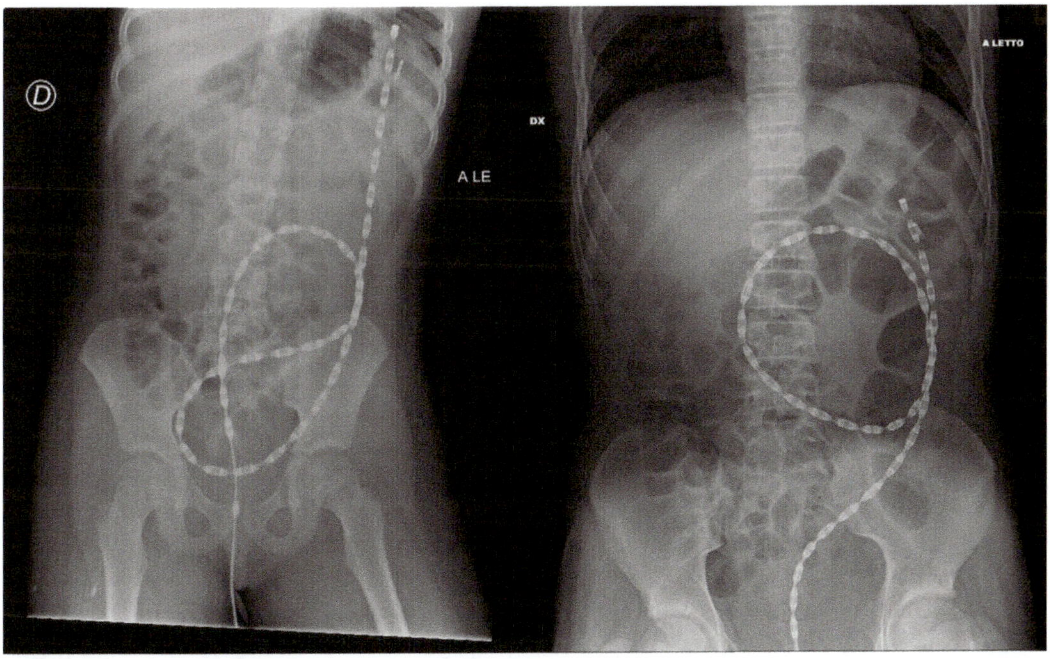

Fig. 7.4 Plain abdominal radiograph showing placement of manometry catheters

7.2.2 Analysis of Colonic Manometry Testing

7.2.2.1 The Colonic Motor Patterns

As for other motility testings, colonic manometry has never been performed in healthy children for ethical reasons; therefore, patterns of normal colon motility have been established, based on the data from children with expected normal colonic physiology and from adult's experience [15].

Colonic motor activity is not rhythmic but is characterized by phasic or brief contractions and tonic or sustained contractions. Different patterns of human colonic phasic pressure activity have been identified by colonic manometry [16, 17]. Traditionally, the motor activity has been classified into propagating or non-propagating, with the latter consisting of apparently random pressure waves recorded at single or multiple recording sites alternated with phases of motor quiescence [18]. However, the advent of HRM has enabled a more detailed characterization of the colonic motor activity, demonstrating that these apparent non-propagating pressure waves actually consist of a series of short-expand single antegrade and retrograde propagating pressure waves. This activity, accounting for most manometry recording, is characterized by a sequence of propagating contractions traveling prominently in a retrograde direction and has been termed cyclic propagating motor pattern. These motor patterns occur at a slow-wave frequency (2–6 cycles per minute) that increase in both number and amplitude in response to a high caloric meal, especially within the distal colon. It has been postulated that this cyclic retrograde activity may serve to hold content within the colon, prevent premature rectal filling, and help control defecation until stool expulsion is socially convenient [17]. In adult patients with slow-transit constipation, the meal fails to increase this distal colonic cyclic activity, potentially indicating an abnormality in the colon's extrinsic neural innervation [19].

Phasic activity increases within a few minutes after a meal and continues for up to 2.5 h depending on the nutrient composition and caloric content of the meal. The presence of a gastrocolic response to a meal, characterized by more frequent and higher amplitude segmental contractions and increased colonic tone, can be calculated by the manometric software as a motility index [4, 5].

Despite making up <2% of the identifiable propagating motor patterns, high-amplitude propagated contractions (HAPCs) constitute the most distinctive and well-studied pattern of colonic motor activity [17, 20]. HAPCs have been considered the most useful marker to assess colonic neuromuscular integrity in clinical practice [4, 5, 20].

The generally accepted definition for HAPCs in children is contractions with an amplitude of greater than 75 mmHg (50–116 mmHg in some studies), a duration of greater than 10 s, and a propagation distance of at least 30 cm [2, 5, 20]. They are expected to stop at the junction between the sigmoid colon and the rectum and are generally associated with the anal sphincter relaxation (coloanal reflex) [21, 22]. HAPCs typically occur following meals and upon awakening, in response to pharmacological stimulation (bisacodyl, neostigmine, or other colonic irritants) or colonic distention [2, 20]. Spontaneous HAPCs arise approximately six times (range 2–24) daily and often precede defecation; they are more frequent in young children and infants and in patients who have had a distal colonic resection [4, 15, 20, 23].

In healthy control, most HAPCs originate in the proximal colon and extend into the distal colon. The primary indicators of abnormal colonic motor function in constipation are reduced HAPC frequency, diminished or absent response to eating, and lack of full propagation [4, 5, 20].

Constipated patients have an overall diminished frequency of HAPCs, but it has also been observed that in some of them, HAPCs normally originate in the proximal colon but fail to progress beyond the transverse colon or are absent [2–5, 16, 24, 25]. Therefore, HAPCs are primarily classified based on their distance of propagation into (1) "fully propagating" when they reach the sigmoid colon (Fig. 7.5); (2) "partially propagating" when they do not propagate beyond the splenic flexure or the descending colon (Fig. 7.6); and (3) "absent" when no

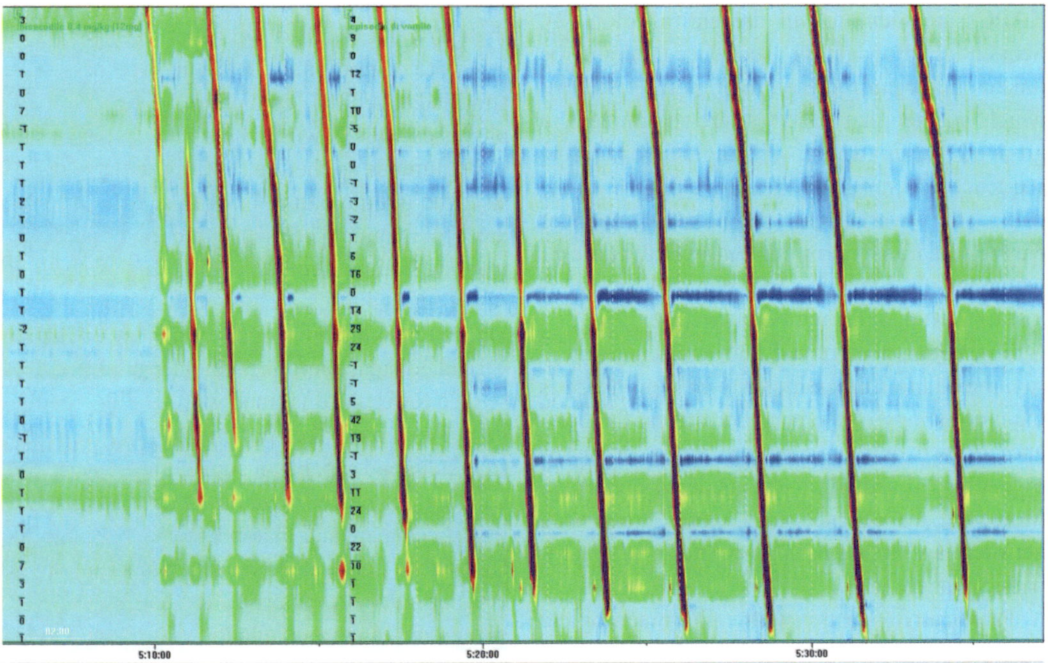

Fig. 7.5 Cluster of fully propagated high-amplitude propagating contractions (HAPCs) recorded with a solid-state catheter (36 pressure recording sites) displayed as isobaric pressure topography plot

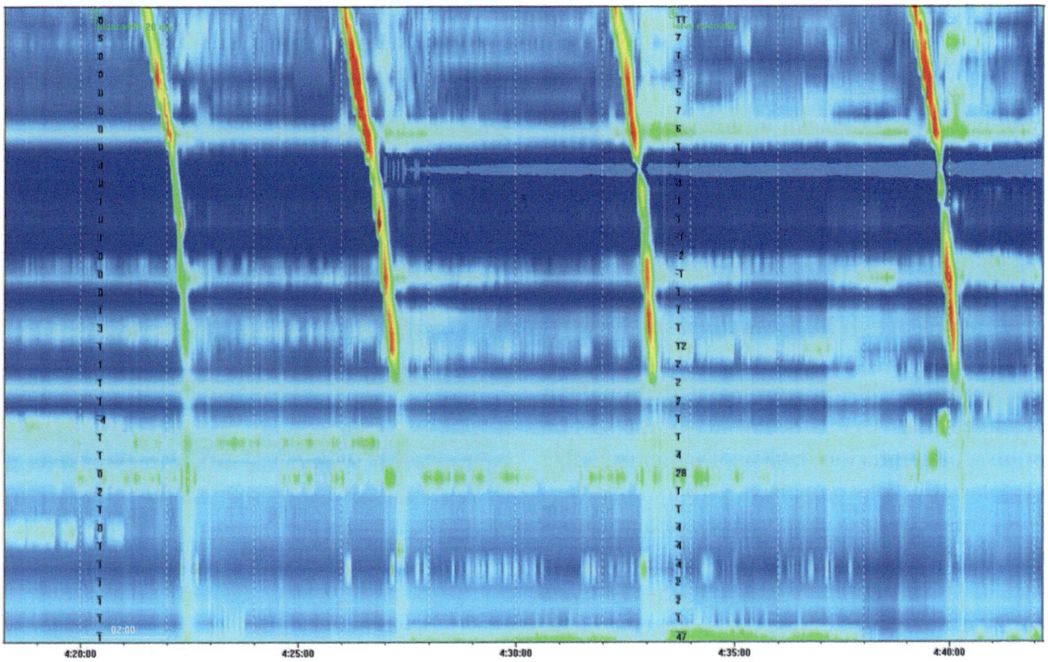

Fig. 7.6 Partially propagating HAPCs: normal propagating contractions in the descending colon and abnormal non-propagating very low-amplitude pressurization in the sigmoid colon

HAPCs are identified in the entire colon [2]. Sequential propagating sequences might be linked in organized spatiotemporal patterns, which means that HAPCs are organized in grouped series where, if single propagating sequences do not span the entire length of the colon, the overlapping of linked sequences ensures the transport of content over the colon's whole length (Fig. 7.7). This regional linkage is lost in most patients with constipation [26].

To evaluate the colonic neuromuscular integrity, the additional parameter to be considered is the morphology of the pressure waves: the presence of two or more pressure peaks or waves exceeding 30 s should be considered abnormal. Moreover, the absence of a motor quiescence between HAPSs has been associated with neuropathic changes [27].

Provocative tests with pharmacological stimuli are an integral part of the study for assessing colonic neuromuscular function integrity [2–5, 28–30]. It is usually performed with bisacodyl infused into the proximal colon at a dose of 0.2 mg/kg of bisacodyl (max 10 mg); however, a doubled dosage (0.4 mg/kg, max 20 mg) has been found to result in a significantly increased number of HAPCs, to improve their propagation toward the distal colon, and to sharpen the morphology of pressure waves (Fig. 7.8). Therefore higher doses of bisacodyl seem to be more accurate in assessing the residual function of colonic motor activity with a possible significant influence in decision-making [31]. Recently a multicenter retrospective study involving children evaluated for treatment-refractory constipation demonstrated that HAPCs occurred within 12 min of bisacodyl infusion in 93% of patients [32].

A less vigorous propulsive activity characterizes low-amplitude propagating contractions (LAPCs); they are defined as having an amplitude of less than 50 mmHg; occur 45–120 times per 24 h, more often during the day than at night; and, as for HAPCs, increase in frequency after meals and upon waking [6]. The function of LAPCs is poorly understood; they are likely to be involved with the transport of less viscous contents, such as liquids and gas [33].

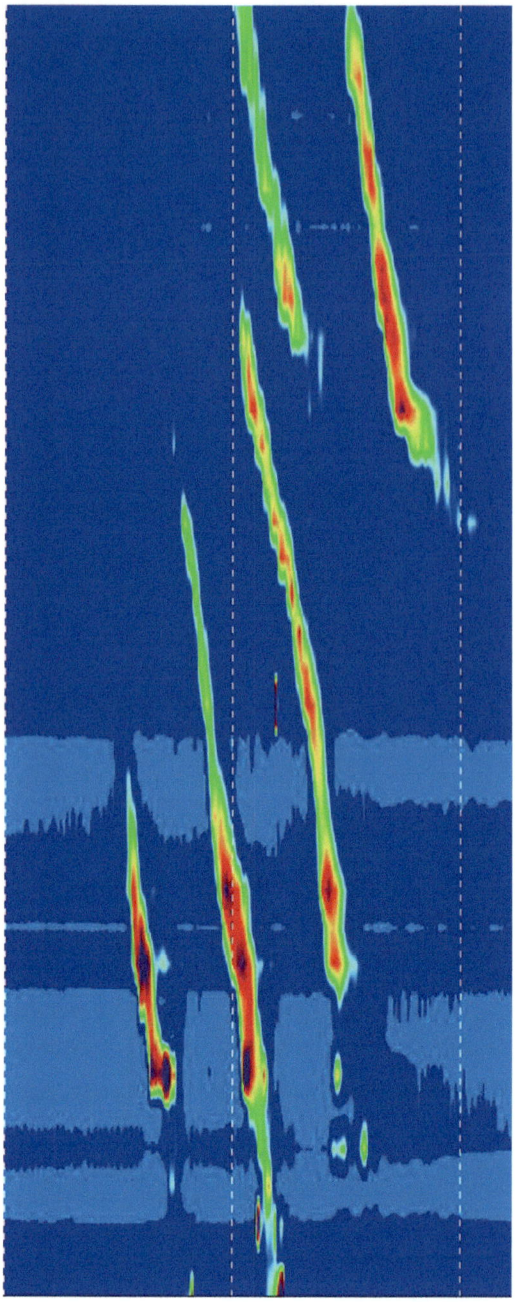

Fig. 7.7 Cluster of overlapping of linked sequences of HAPCs

7.2.2.2 Interpretation and Indications

In clinical practice, colonic manometry has been proved to impact the management of constipation in children significantly. It is recommended to perform colonic manometry in patients with

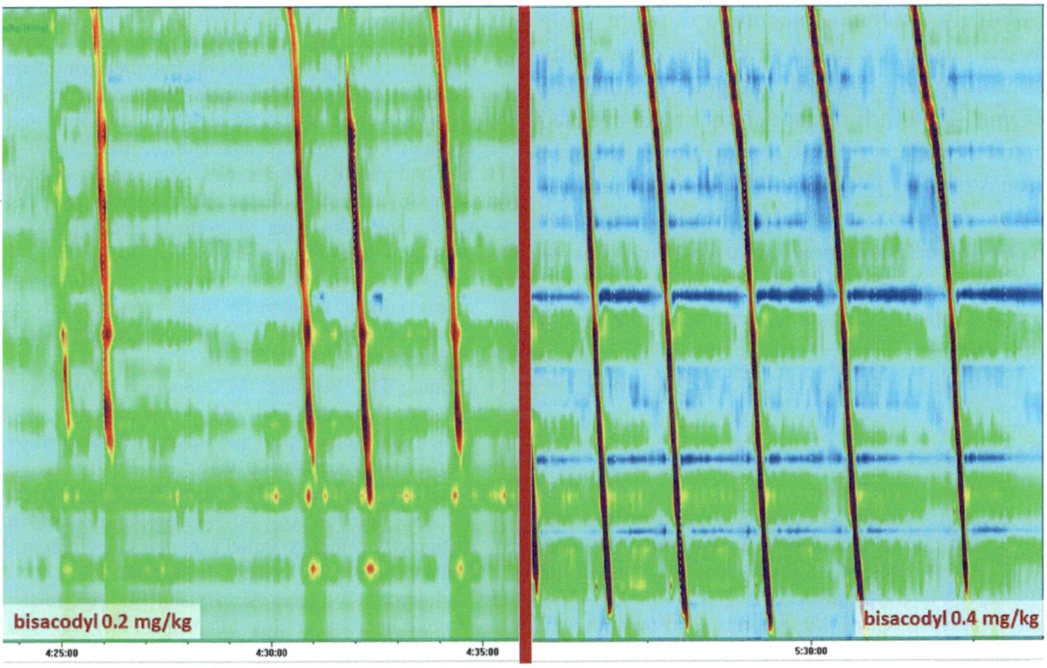

Fig. 7.8 Tracing of the same patient showing the post-bisacodyl motor activity after the dose of 0.2 mg/kg and 0.4 mg/kg of bisacodyl: the improvement in number, propagation toward the colon, and morphology of pressure waves is clearly identifiable

treatment-refractory constipation and with evidence of slow colonic transit in the absence of an evacuatory disorder before considering surgical intervention.

There is an agreement that a study can be considered normal when there is an increase in motility after a meal and the occurrence of spontaneous, meal-induced, or bisacodyl-induced HAPC propagating to the rectosigmoid junction [4, 5].

Colon manometry may differentiate children with functional constipation from those with a colonic motor disorder, with the former showing normal colonic motility with an intact gastrocolic response to a meal [25]. Colonic motor dysfunction's identification and location may provide useful information for planning treatment and predicting therapeutic outcomes in children in which conventional medical and behavioral treatments have failed [5, 29, 34].

Reduced HAPC frequency and a diminished or absent response to eating or pharmacological stimulation indicate abnormal colonic motility. Suppressed response to meal ingestion indicates colonic myopathy and an absent response of a possible neuropathy [16]. The failed response to pharmacological stimuli indicates abnormality within the myenteric plexus, cholinergic pathways, or recto-colonic neural pathways [35–37].

In selected patients with constipation refractory to medical therapy, colonic manometry offers useful information to planning surgical interventions. Children with no identifiable HAPCs are less likely to benefit from using antegrade enemas via cecostomy, while a normal colonic response to bisacodyl predicts a favorable outcome [38, 39]. Colon manometry is deemed useful also in guiding partial colonic resection. Indeed some authors reported that colonic segments with abnormal motility on colonic manometry might benefit from surgical resection [39–41]. A recent study found that although not predictive of successful medical intervention, an abnormal colonic manometry, mostly the presence of partially propagated HAPCs, is predictive of surgery in children [42].

Finally, colonic manometry might help assess the improvement of colonic dysmotility after long-term use of antegrade continence enema [43, 44]. It might determine the diverted colon's function prior to a possible resection or reanastomosis [39]. Finally, colonic manometry is deemed useful in guiding the management of persistent postoperative symptoms of patients with Hirschsprung's disease and anorectal malformations (e.g., imperforate anus) [45, 46].

References

1. Bleier JIS, Wilkins KB. Colonic physiology. In: Steele S, Hull T, Read T, Saclarides T, Senagore A, Whitlow C, editors. The ASCRS textbook of colon and Rectal surgery. Cham: Springer; 2016. p. P27–35.
2. Corsetti M, Costa M, Bassotti G, Bharucha AE, Borrelli O, Dinning P, et al. First translational consensus on terminology and definitions of colonic motility in animals and humans studied by manometric and other techniques. Nat Rev Gastroenterol Hepatol. 2019;16:559–79.
3. Tambucci R, Quitadamo P, Thapar N, Zenzeri L, Caldaro T, Staiano A, et al. Diagnostic tests in pediatric constipation. J Pediatr Gastroenterol Nutr. 2018;66:e89–98.
4. Camilleri M, Bharucha AE, di Lorenzo C, Hasler WL, Prather CM, Rao SS, Wald A. American Neurogastroenterology and Motility Society consensus statement on intraluminal measurement of gastrointestinal and colonic motility in clinical practice. Neurogastroenterol Motil. 2008;20:1269–82.
5. Rodriguez L, Sood M, Di Lorenzo C, Saps M. An ANMS-NASPGHAN consensus document on anorectal and colonic manometry in children. Neurogastroenterol Motil. 2017;29
6. Rao SS, Sadeghi P, Beaty J, Kavlock R, Ackerson K. Ambulatory 24-h colonic manometry in healthy humans. Am J Physiol Gastrointest Liver Physiol. 2001;280:629–39.
7. Dinning PG, Carrington EV, Scott SM. The use of colonic and anorectal high-resolution manometry and its place in clinical work and in research. Neurogastroenterol Motil. 2015;27:1693–708.
8. Tabbers MM, Di Lorenzo C, Berger MY, Faure C, Langendam MW, Nurko S, et al. Evaluation and treatment of functional constipation in infants and children: evidence-based recommendations from ESPGHAN and NASPGHAN. J Pediatr Gastroenterol Nutr. 2014;58:258–74.
9. Bharucha AE, Pemberton JH, Locke GR 3rd. American Gastroenterological Association technical review on constipation. Gastroenterology. 2013;144:218–38.
10. Dinning PG, Wiklendt L, Gibbins I, Patton V, Bampton P, Lubowski DZ, et al. Low-resolution colonic manometry leads to a gross misinterpretation of the frequency and polarity of propagating sequences: initial results from fiber-optic high-resolution manometry studies. Neurogastroenterol Motil. 2013;25:640–9.
11. El-Chammas KI, Tipnis NA, Simpson PM, Sood MR. Colon high-resolution manometry: using pressure topography plots to evaluate pediatric colon motility. J Pediatr Gastroenterol Nutr. 2014;59:500–4.
12. Lee YY, Erdogan A, Rao SS. How to perform and assess colonic manometry and barostat study in chronic constipation. J Neurogastroenterol Motil. 2014;20:547–52.
13. Dinning PG, Benninga MA, Southwell BR, Scott SM. Paediatric and adult colonic manometry: a tool to help unravel the pathophysiology of constipation. World J Gastroenterol. 2010;16:5162–72.
14. Sanghavi RM, Centeno RM, Barth B. Single balloon-assisted colonoscopy for placement of colonic manometry catheters: initial experience in children. J Pediatr Gastroenterol Nutr. 2018;67:194–7.
15. Di Lorenzo C, Flores AF, Hyman PE. Age-related changes in colon motility. J Pediatr. 1995;127:593–6.
16. Rao SS, Sadeghi P, Beaty J, Kavlock R. Ambulatory 24-hour colonic manometry in slow-transit constipation. Am J Gastroenterol. 2004;99:2405–16.
17. Dinning PG, Wiklendt L, Maslen L, Gibbins I, Patton V, Arkwright JW, et al. Quantification of in vivo colonic motor patterns in healthy humans before and after a meal revealed by high-resolution fiber-optic manometry. Neurogastroenterol Motil. 2014;26:1443–57.
18. Dinning PG, Szczesniak MM, Cook IJ. Proximal colonic propagating pressure waves sequences and their relationship with movements of content in the proximal human colon. Neurogastroenterol Motil. 2008;20:512–20.
19. Dinning PG, Wiklendt L, Maslen L, Patton V, Lewis H, Arkwright JW, et al. Colonic motor abnormalities in slow transit constipation defined by high resolution, fibre-optic manometry. Neurogastroenterol Motil. 2015;27:379–88.
20. Bharucha AE. High amplitude propagated contractions. Neurogastroenterol Motil. 2012;24:977–82.
21. Rodriguez L, Siddiqui A, Nurko S. Internal anal sphincter relaxation associated with bisacodyl-induced colonic high amplitude propagating contractions in children with constipation: a Colo-anal reflex? Neurogastroenterol Motil. 2012;24:1023–e545.
22. Sintusek P, Rybak A, Mutalib M, Thapar N, Borrelli O, Lindley KJ. Preservation of the Colo-anal reflex in colonic transection and postoperative Hirschsprung's disease: potential extrinsic neural pathway. Neurogastroenterol Motil. 2019;31:e13472.
23. Kaul A, Garza JM, Connor FL, Cocjin JT, Flores AF, Hyman PE, et al. Colonic hyperactivity results in frequent fecal soiling in a subset of children after surgery for Hirschsprung disease. J Pediatr Gastroenterol Nutr. 2011;52:433–6.

24. Hagger R, Kumar D, Benson M, Grundy A. Colonic motor activity in slow-transit idiopathic constipation as identified by 24-h pancolonic ambulatory manometry. Neurogastroenterol Mot. 2003;15:515–22.
25. Di Lorenzo C, Flores AF, Reddy SN, Hyman PE. Use of colonic manometry to differentiate causes of intractable constipation in children. J Pediatr. 1992;120:690–5.
26. Dinning PG, Zarate N, Hunt LM, Fuentealba SE, Mohammed SD, Szczesniak MM, et al. Pancolonic spatiotemporal mapping reveals regional deficiencies in, and disorganization of colonic propagating pressure waves in severe constipation. Neurogastroenterol Motil. 2010;22:340–9.
27. Giorgio V, Borrelli O, Smith VV, Rampling D, Köglmeier J, Shah N, et al. High-resolution colonic manometry accurately predicts colonic neuromuscular pathological phenotype in pediatric slow transit constipation. Neurogastroenterol Motil. 2013;25:70–8.
28. Dinning PG, Di Lorenzo C. Colonic dysmotility in constipation. Best Pract Res Clin Gastroenterol. 2011;25:89–101.
29. Pensabene L, Youssef NN, Griffiths JM, Di Lorenzo C. Colonic manometry in children with defecatory disorders. Role in diagnosis and management. Am J Gastroenterol. 2003;98:1052–7.
30. Hamid SA, Di Lorenzo C, Reddy SN, Flores AF, Hyman PE. Bisacodyl and high-amplitude-propagating colonic contractions in children. J Pediatr Gastroenterol Nutr. 1998;27:398–402.
31. Borrelli O, Pescarin M, Saliakellis E, Tambucci R, Quitadamo P, Valitutti F, et al. Sequential incremental doses of bisacodyl increase the diagnostic accuracy of colonic manometry. Neurogastroenterol Motil. 2016;28:1747–55.
32. Heitmann PT, Wiklendt L, Thapar N, Borrelli O, Di Lorenzo C, Yacob DT, et al. Characterization of the colonic response to bisacodyl in children with treatment-refractory constipation. Neurogastroenterol Motil. 2020;32:e13851.
33. Bassotti G, Clementi M, Antonelli E, Pelli MA, Tonini M. Low amplitude propagated contractile waves: a relevant propulsive mechanism of human colon. Dig Liver Dis. 2001;33:36–40.
34. Martin MJ, Steele SR, Mullenix PS, Noel JM, Weichmann D, Azarow KS. A pilot study using total colonic manometry in the surgical evaluation of pediatric functional colonic obstruction. J Pediatr Surg. 2004;39:352–9.
35. Bassotti G, Chiarioni G, Imbimbo BP, Betti C, Bonfante F, Vantini I, et al. Impaired colonic motor response to cholinergic stimulation in patients with severe chronic idiopathic (slow transit type) constipation. Dig Dis Sci. 1993;38:1040–5.
36. Preston DM, Lennard-Jones JE. Pelvic motility and response to intraluminal bisacodyl in slow-transit constipation. Dig Dis Sci. 1985;30:289–94.
37. Dinning PG, Bampton PA, Kennedy ML, Lubowski DZ, King D, Cook IJ. Impaired proximal colonic motor response to rectal mechanical and chemical stimulation in obstructed defecation. Dis Colon Rectum. 2005;48:1777–84.
38. Van den Berg MM, Hogan M, Caniano DA, Di Lorenzo C, Benninga MA, Mousa HM. Colonic manometry as predictor of cecostomy success in children with defecation disorders. J Pediatr Surg. 2006;41:730–6.
39. Villarreal J, Sood M, Zangen T, Flores A, Michel R, Reddy N, et al. Colonic diversion for intractable constipation in children: colonic manometry helps guide clinical decisions. J Pediatr Gastroenterol Nutr. 2001;33:588–91.
40. Gasior A, Reck C, Vilanova-Sanchez A, Diefenbach KA, Yacob D, Lu P, et al. Surgical management of functional constipation: an intermediate report of a new approach using a laparoscopic sigmoid resection combined with Malone appendicostomy. J Pediatr Surg. 2018;53:1160–2.
41. Gupta A, Basson S, Borrelli O, Lindley K, Thapar N, Saliakellis E, et al. Surgically treated intractable constipation in children evaluated with colonic manometry. J Pediatr Surg. 2020;55:265–8.
42. Rodriguez L, Heinz N, Nurko S. Utility of colon manometry in guiding therapy and predicting need for surgery in children with defecation disorders. J Pediatr Gastroenterol Nutr. 2020;70:232–7.
43. Rodriguez L, Nurko S, Flores A. Factors associated with successful decrease and discontinuation of antegrade continence enemas (ACE) in children with defecation disorders: a study evaluating the effect of ACE on colon motility. Neurogastroenterol Motil. 2013;25:140–e81.
44. Aspirot A, Fernandez S, Di Lorenzo C, Skaggs B, Mousa H. Antegrade enemas for defecation disorders: do they improve the colonic motility? J Pediatr Surg. 2009;44:1575–80.
45. Di Lorenzo C, Solzi GF, Flores AF, Schwankovsky L, Hyman PE. Colonic motility after surgery for Hirschsprung's disease. Am J Gastroenterol. 2000;95:1759–64.
46. Heikenen JB, Werlin SL, Di Lorenzo C, Hyman PE, Cocjin J, Flores AF, et al. Colonic motility in children with repaired imperforate anus. Dig Dis Sci. 1999;44:1288–92.

8 Anorectal Malformations: Anorectal Manometric and Endosonographic Combined Approach

Tamara Caldaro

8.1 Introduction

Assessment and management of post-operative problems following surgical repair of anorectal malformations (ARMs) represent a specific challenge for paediatric surgeons and paediatric gastroenterologists due to the complexity of the disease.

Despite technical advances and accurate anatomical reconstructions, high prevalence of faecal incontinence (16.7%–76.7%) and chronic constipation (22.6%–86.7%) has been reported in the long-term follow-up, without clear differences in the rate of impaired bowel movements in patients with low or high ARMs [1].

These types of disabilities have a severe impact on patients' quality of life, resulting in many psychologic and social disturbances [2]. Several studies have investigated a wealth of clinical information and various possible prognostic factors that could be related to the outcome [3].

Based on three anatomical elements (type of ARMs, spinal defects and sacral anomalies), an "ARM Continence Predictor Index" has been created to define a personalised potential continence score system prior to surgical intervention, so as to plan a tailored bowel management programme to successfully get patients clean [4].

T. Caldaro (✉)
Digestive Surgery and Endoscopy Unit, Bambino Gesù Children's Hospital-IRCCS, Rome, Italy
e-mail: tamara.caldaro@opbg.net

However, a comprehensive assessment of pelvic floor is necessary in order to better define the appropriate treatment.

Post-operative investigations, that include endoanal ultrasonography (EAUS) and anorectal manometry, provide accurate objective information about the continence status, as they analyse the global anorectal anatomy and functionality [5].

8.2 Anorectal Malformations and Anorectal Dysfunction

ARMs include a wide selection of congenital anomalies that range from the most benign of all defects, the perineal fistula, to the severe anomalies, such as cloaca and cloacal exstrophy [6]; consequently, various anatomical classification systems have been devised to define the pathology of these anorectal anomalies. In 1970, an international classification was proposed for both genders, dividing the anomalies into high, intermediate and low, according to the position of the terminal rectum to the levator ani. In 1984, the Wingspread classification distinguished between high (supra-levator), intermediate (partially trans-levator) and low (trans-levator) ARMs in male and female subjects, with the introduction of special groups for cloacal and rare malformations [7].

Later, Peña proposed a surgically oriented classification based on the presence and position

M. Lima, G. Ruggeri (eds.), *Ano-Rectal Endosonography and Manometry in Paediatrics*,
https://doi.org/10.1007/978-3-030-97668-2_8

of fistula and on the relationship of the terminal colon to the levator sling muscles of the pelvic floor [8].

The location and type of fistula help to guide the operative approach and to determine the extent of mobilisation needed for pulling through the blind pouch. This classification system was also the first one which attempted to define prognosis for each group in terms of functional bowel outcomes. In fact, even after the introduction of the posterior sagittal anorectoplasty (PSARP), the major post-operative problems observed were chronic constipation and overflow incontinence caused by motility disturbances and stool incontinence secondary to true sphincter insufficiency.

In 2005, the Krickenbeck group suggested a new classification to rationalise functional outcome among different studies and to allow more meaningful comparisons. Consequently, the classification comprises three distinct categories: diagnostic, surgical procedure and post-operative functional results [9, 10]. The anorectal anomalies are divided into two groups, the major clinical groups and the rare/regional variants, based on the presence, type and location of fistula, as well as the position of the rectal pouch. For follow-up studies, not only the site of the fistula should be documented but also other additional groupings considering specifically the operative procedure performed and the main post-operative motility disturbances (voluntary bowel movements, soiling and constipation).

Difficulty in finding an exhaustive classification comes from the complexity and variety of the disease, whose prognosis also depends on associated anomalies, which could influence the patients' quality of life.

Normal control of defaecation requires integrity of neuromuscular structures including rectum, internal and external sphincters and pelvic floor muscles; preservation of anorectal sensation; maintenance of neural (both autonomic and somatic) pathways; and normal intestinal transit time [11].

Commonly, one or more of these mechanisms may be altered in patients with ARMs.

As regards the sphincter mechanisms, internal anal sphincter (IAS) and external anal sphincter (EAS) may be practically normal in minor defects but very compromised in the more complex ones, with severe degrees of muscle underdevelopment and atrophy. In addition to these congenital defects, scars of the anorectal sphincters are frequently detected after surgical reconstruction [12].

Megarectum, a condition found in 10–50% of patients with ARMs, is responsible for constipation and overflow incontinence due to accumulation of stool in the dilated segment [13].

Various aetiologies of primary megarectum in ARMs have been proposed, including sacral neuropathy leading to loss of proprioception and denervation [14, 15] or in utero caudal obstruction of the hindgut during development leading to a grossly dilated rectosigmoid. Post-surgical anorectoplasty denervation, chronic constipation and anastomotic anal stricture are the suggested causes of secondary megarectum [16, 17].

The dilated sigmoid and rectal segment are characterised by a delay in the transit time (hypomotility) and a higher sensory threshold, conditions that are more critical in the presence of sacral and/or spinal anomalies.

The anorectal reconstruction is another crucial step in preserving faecal continence.

In reconstructive ARMs surgery, the PSARP procedure is broadly accepted for its potential advantages, but although it provides the most accurate anatomical reconstruction saving the anal sphincters, the outcomes are not in keeping with the technical benefit. The aims of the technique are to place the rectum in the middle of the anal sphincter complex, which is accurately mapped with an electrical stimulator, and to avoid injuries to adjacent sphincter structures [18].

However, mislocation of the anoplasty is a frequently seen complication. The anus can be mislocated anterior or posterior to the muscle complex or in a lateral orientation. Location of the anoplasty outside of the centre of the anal sphincters can impact the ability to voluntarily handle bowel movements and should be corrected if the potential for bowel control is present [6].

Laparoscopic-assisted anorectoplasty (LAARP) may be preferable to the posterior sagittal method to define the right placement of the anus in high anomalies (recto-bladder neck fistula and complex cloacae), which require an abdominal approach.

LAARP provides direct visualisation of the rectal fistula and surrounding structures, identifies the central portion of the puborectalis from inside reducing the insurgence of injury to nerve plexus and allows accurate placement of the bowel through the anatomical midline and the levator sling with minimal surgical trauma to the continence mechanism [19].

The overall voluntary bowel movement, constipation and faecal incontinence in the series on PSARP by Peña et al. [20] were 75%, 48% and 25%, respectively, while LAARP analysis by Pathak et al. [21] displayed 81% voluntary bowel movement, 9% constipation and 12% faecal incontinence.

In literature, however, the pre-eminence of LAARP over PSARP is not clearly demonstrated, due to lack of homogeneous data and identical criteria regarding the functional outcome assessment and the long-term follow-up [21].

Therefore, accurate evaluation of patients with ARMs remains a challenge in consideration of the complexity of congenital and acquired factors which potentially influence the mechanism of faecal continence, even after proper surgical reconstruction. A directed history and a careful physical examination with particular care for the integrity of the perineum and rectum, a neurologic evaluation and a complete diagnostic assessment are mandatory to define a bowel management programme that can improve the quality of life of patients.

8.3 Anorectal Manometry

Anorectal manometry is a recognised procedure used in the full and proper assessment of defaecatory disorders and also in the pre- and/or post-operative evaluation of the anorectal area.

The ANMS-NASPGHAN consensus document on anorectal and colonic manometry in children recommends to perform anorectal manometry on patients treated surgically for ARMs with persistent defaecation problems [22].

Recent advances in diagnostic techniques have led to an evolution in classical anorectal manometry test with the introduction of both the high-resolution anorectal manometry (HR-ARM) and the three-dimensional high-definition anorectal manometry (3D-HDARM) [23].

Conventional manometry is carried out using a sleeve catheter, a water-perfused and a data acquisition system. The manometry probes have 4–8 side holes that are disposed helicoidally or radially along the catheter and connected to the perfusion apparatus with a pneumohydraulic pump. Data are presented as pressure lines. The catheter is incapable of acquiring the pressures of the entire anal canal simultaneously. Therefore, a pull-through manoeuvre is required to sample the entire area of interest.

In HR-ARM, the recorded data are displayed as highly detailed topographical colour-contoured plots, rather than overlapping line traces. With the improvement of electronics, new manometric water-perfused and solid-state catheters with many miniature pressure sensors have been developed in order to ensure a better interpretation of pressure changes and to allow a simultaneous assessment of the anal sphincters. These new probes are also useful in minimising the motion artefacts and eliminating the pull-through manoeuvres [24].

HDARM allows for three-dimensional physiologic mapping of anal sphincters with the possibility of detailed assessment of pressures and visualisation of symmetry by means of a single introduction of the probe. In fact, thanks to the high number of sensors placed circumferentially on the catheter, 3D pressure models of the organ can be obtained, with a better definition of the different components of the anal canal, including the puborectalis muscle, the EAS and the IAS. Anyway, the anal resting pressure may result overestimated and the canal dynamics upon balloon distension can be misinterpreted with the current catheter, as it has an outer diameter of almost 11 mm and according to the Laplace's law, which states that pressure within a tube is inversely proportional to the radius [25].

Before performing anorectal manometry, especially in patients with ARMs, it is recommended to:

- Classify the type of ARMs and assess associated sacral and/or spinal anomalies
- Inspect carefully the anorectal region and the anal opening to exclude anal stricture or pronounced opening, rectal mucosa prolapse or mislocation of the neo-anus
- Evaluate the presence of a megarectum
- Ensure a complete rectal disimpaction

In fact, any of these factors could modify the results of manometric tests and in some conditions (e.g. anal stricture) the exam could be contraindicated.

8.4 Manometric Parameters and Anorectal Malformations

The manometric test provides comprehensive information regarding (**a**) the pressures of the anal sphincter muscles at rest and during dynamic manoeuvres (squeezing and straining); (**b**) the rectal sensation; (**c**) the innervation of IAS and EAS eliciting the recto-anal inhibitory reflex (RAIR) and the cough reflex; (**d**) the defaecation dynamics (bear-down manoeuvre) by assessing the recto-anal pressure gradient during straining and the balloon expulsion test (BET).

8.4.1 Anal Sphincter Pressures

Function and integrity of the anal sphincters are evaluated by measuring the pressure at different levels of the anal canal. Modifications of pressure values could be useful to determine the location and grade of anal sphincter defects.

A low anal resting pressure (ARP) may identify an underlying problem with the IAS, which supplies between 55% and 85% of resting pressure, whereas a decrease in voluntary anal squeeze pressure (ASP) could be related to a defect of the EAS.

As numerous studies have detected, lower ARP and ASP values were found in patients with severe types of ARMs and/or in those with faecal incontinence due to wide sphincter defects or atrophy [26, 27].

8.4.2 Rectal Sensation

Using graded balloon, rectal sensation can be measured by assessing three sensory thresholds: first sensation, urge to evacuate and discomfort.

In case of megarectum, a frequent condition in ARMs, manometry testing shows an impaired defaecatory sensation (rectal hyposensitivity) with an increase of 1 or more of the threshold values, because higher volumes of distention are needed to feel the defaecatory stimulus. Consequently, constipation and soiling appear to be more common among patients with a large rectal volume (>150 mL of air), as Hedlund et al. documented [26].

8.4.3 Reflexes

Innervation of the IAS and the EAS can be studied eliciting the RAIR via the myenteric plexus and the cough reflex via the spinal reflex arc, respectively.

RAIR is a relaxation response in the IAS, namely a pressure drop of at least 25% in the anal canal following rectal distension. It is elicited by rapid insufflation of minimum 20 mL of air into a balloon positioned in the distal rectum at the level of the proximal high-pressure zone.

It is absent in Hirschsprung disease, but also in other conditions such as lower anterior resections and injuries of IAS. Potential technical pitfalls with false-negative results of RAIR could be obtained in the presence of two typical disorders observed in patients operated on ARMs: megarectum, for which a greater volume of rectal distention is required to elicit RAIR, and low ARP due to lesions of the IAS [12, 13].

RAIR is considered an indirect sign of continence. Rintala et al. [28] reported RAIR as positive in all the “good” and in 75% of the “fair” group patients and negative in the "bad" group with ARMs. Same results were achieved by Senel et al. [27].

The "cough reflex" is a manoeuvre indicated to assess the integrity of spinal reflex pathways in patients with incontinence. It consists of a contraction of the EAS following an intra-abdominal pressure increment (e.g. coughing) [29].

The abnormality of innervation of EAS has been identified as one of the most important factors which affect post-operative anal function in patients with ARMs. An altered response may suggest damages in lumbosacral and motor nerves ending in puborectalis and in EAS, even when the lesion is secondary to a congenital anomaly of lumbosacral vertebrae or to a spinal dysraphism.

William et al. reported that 35% of subjects operated on ARMs had lumbosacral vertebral abnormalities on plain X-ray film, whereas 53% showed abnormalities of spinal cord or vertebrae in magnetic resonance imaging (MRI) study [30].

Capitanucci et al., performing urodynamic evaluations on patients with anorectal anomalies as well as spinal dysraphism, demonstrated that the latter may be asymptomatic in small children [31].

Thus, an accurate functional and morphological evaluation of the lumbosacral spine using neurophysiological tests and MRI is essential to ensure an early diagnosis and treatment of the spinal cord lesion before the clinical symptom appears [32].

8.4.4 Bear-Down Manoeuvre and Balloon Expulsion Test

Normal defaecation dynamics is characterised by an abdominal compression associated with anal relaxation. The bear-down manoeuvre simulates evacuation and it is used to assess anorectal and pelvic floor pressure changes during attempted defaecation. Pelvic floor dyssynergia, defined as failing relaxation and coordination of pelvic floor and abdominal muscles during evacuation, is diagnosed if these coordinated movements do not occur.

BET is a helpful screening test to confirm the presence of dyssynergia because it mimics the stool in the rectal vault that should be expelled in 1 min. It is characterised by high specificity (80% to 90%), although the sensitivity is lower than 50% [33]. There are four manometric patterns of pelvic floor dyssynergia according to Rao classification [34, 35]:

- Type I: Adequate rectal push effort with paradoxical anal sphincter contraction.
- Type II: Inadequate rectal push effort with paradoxical anal sphincter contraction.
- Type III: Adequate rectal push effort but inadequate relaxation (<20%) of anal sphincter pressure.
- Type IV: Inadequate rectal push effort and also inadequate relaxation (<20%) of anal sphincter pressure.

A large proportion of patients with chronic constipation suffer from dyssynergic defaecation and the prevalence ranges from 25% to 50% in adult population and up to 50% in children [36, 37]. The incidence is not defined in patients with ARMs; anyway some studies have documented a manometric dyssynergic pattern also in subjects without constipation [38, 39].

The manometric patterns of a patient surgically treated for ARMs and of a child with sacral agenesis plus non-treated anterior anus are shown in Figs. 8.1 and 8.2, respectively.

8.5 Anorectal Manometric Findings: Data from Literature

Since the late 1970s, some authors have used anorectal manometry to estimate anal pressure in patients operated on ARMs.

One of the first studies on patients with intermediate and high ARMs demonstrated that the anorectal pressure profile had no marked high-pressure zone in the anal canal in subjects suffering from faecal incontinence and that only 20% of the patients with high-type anomalies had a RAIR, 23% of these had poor clinical results [40].

a

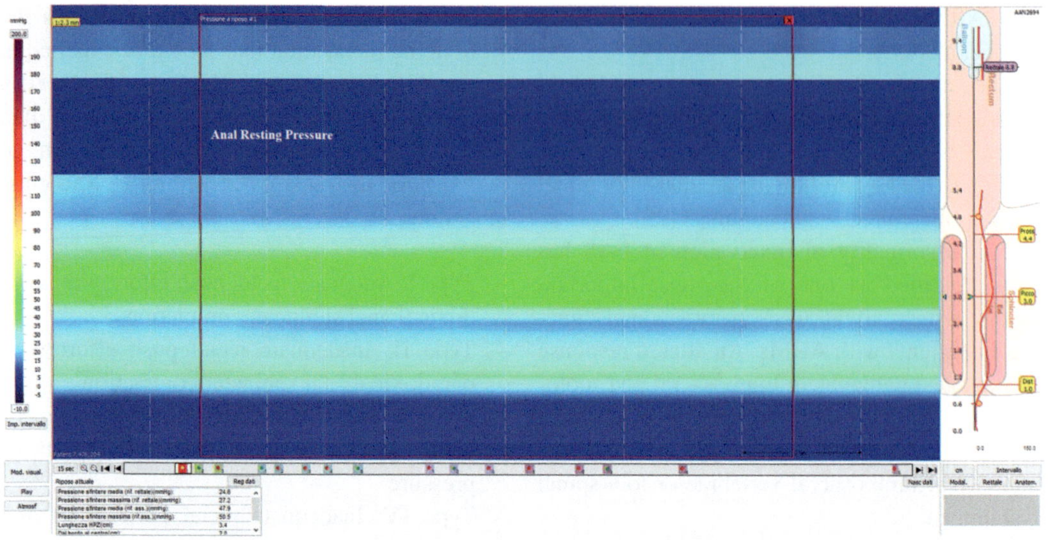

b

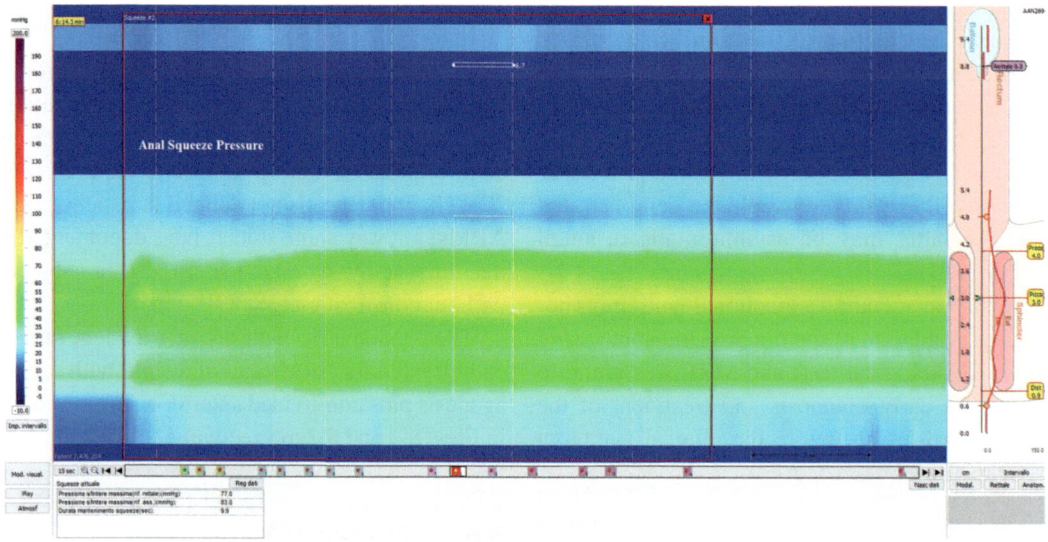

Fig. 8.1 5-year-old female child, anorectal malformation and recto-vestibular fistula, with history of constipation and soiling

HR-ARM documents low ARP (**a**) and ASP (**b**), pelvic floor dyssynergia type I according to Rao's classification (**c**), absence of RAIR (**d**)

HR-ARM High resolution anorectal manometry, *ARP* Anal Resting Pressure, *ASP* Anal Squeeze Pressure, *RAIR* Recto-Anal Inhibitory Reflex

Source: Digestive Surgery and Endoscopy Unit; Bambino Gesù Children's Hospital-IRCCS-Rome, Italy

c

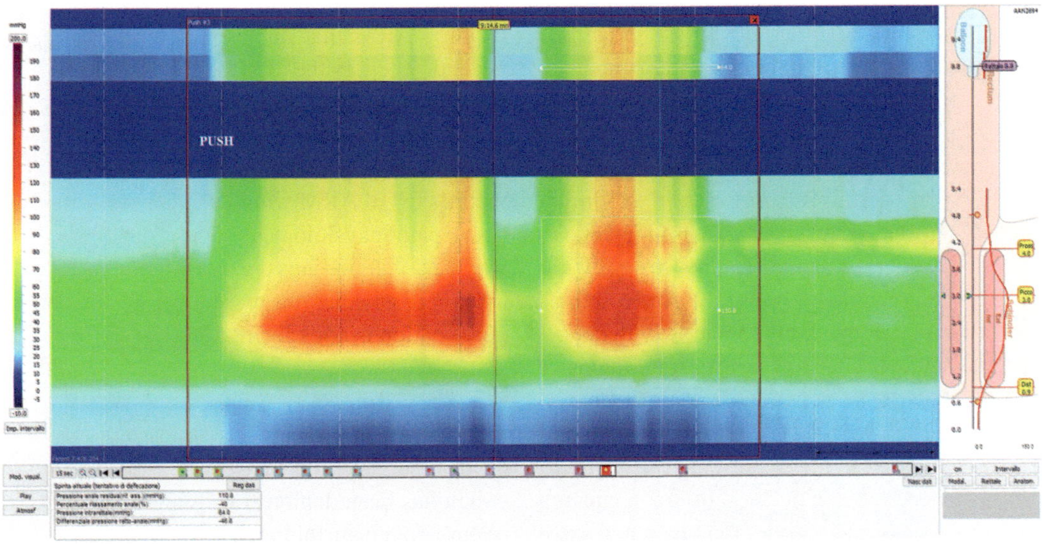

d

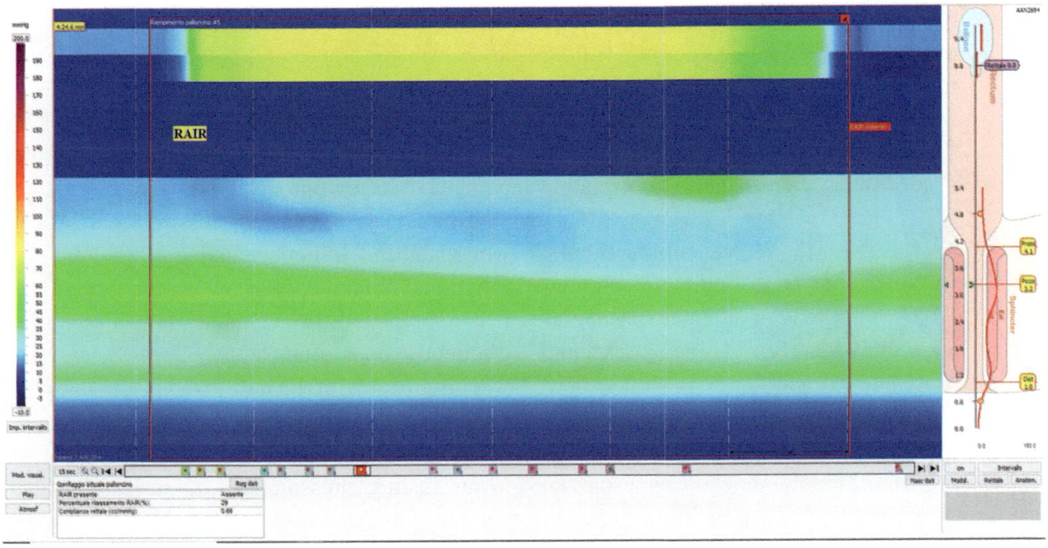

Fig. 8.1 (continued)

In 1998, the same authors reported long-term clinical and manometric data with regard to bowel function in a group of 47 patients treated surgically for ARMs who were 10–30 years.

In comparison to the previous group, poor results were observed in 12% of patients with high-type anomalies, the high-pressure zone was present in 73%, while the RAIR was elicited only in 20% of subjects. Therefore, as a result of these manometric evidences, the presence of RAIR and high-pressure zone was considered to be a chief parameter for objective assessment of anal

a

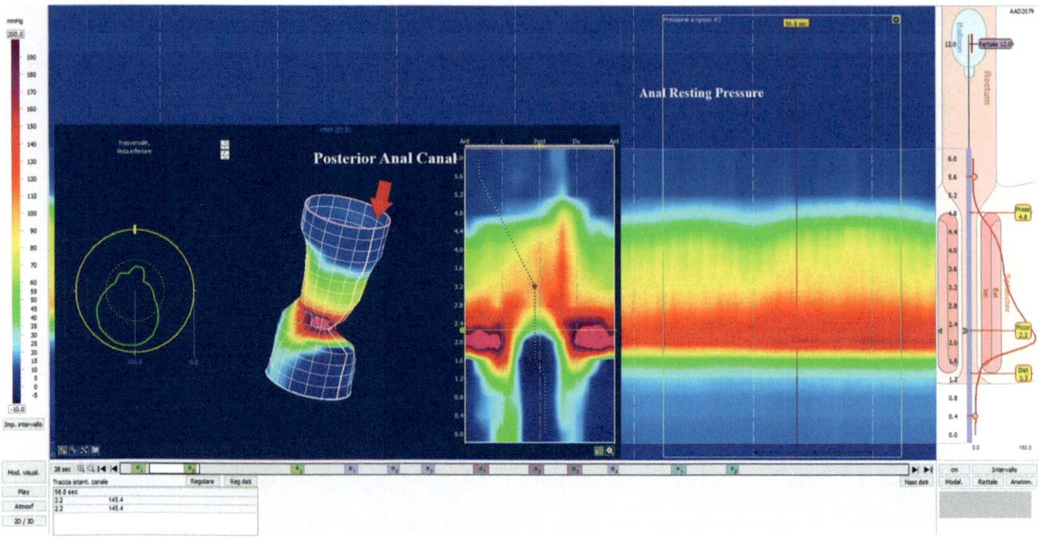

b

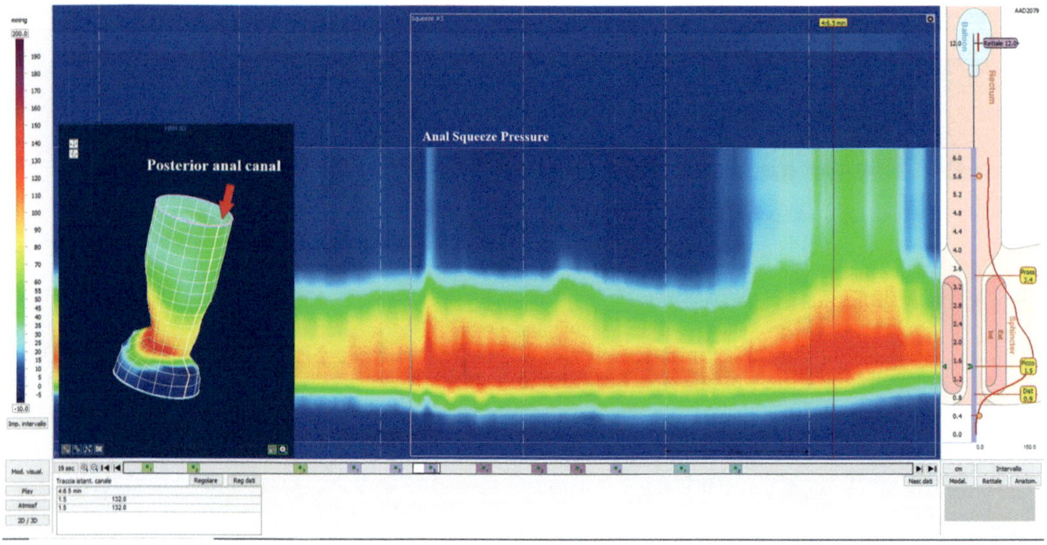

Fig. 8.2 6-year-old female patient with non-treated anterior anus and sacral agenesis, suffering from chronic constipation

HD-ARM detects high ARP (**a**) and ASP (**b**), pelvic floor dyssynergia type I according to Rao's classification (**c**), presence of RAIR (**d**)

HD-ARM High definition anorectal manometry, *ARP* Anal Resting Pressure, *ASP* Anal Squeeze Pressure, *RAIR* Recto-Anal Inhibitory Reflex

Source: Digestive Surgery and Endoscopy Unit; Bambino Gesù Children's Hospital-IRCCS-Rome, Italy

c

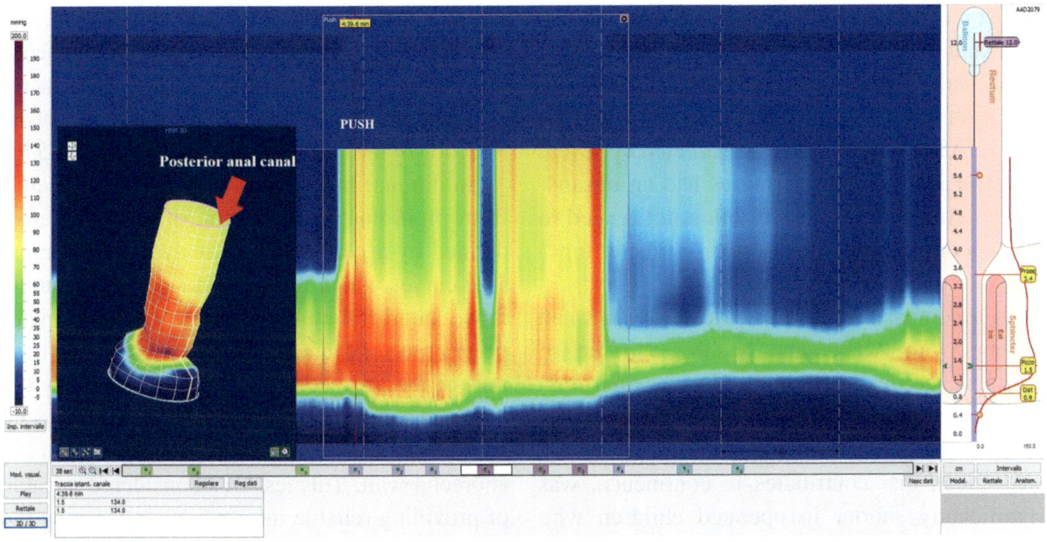

d

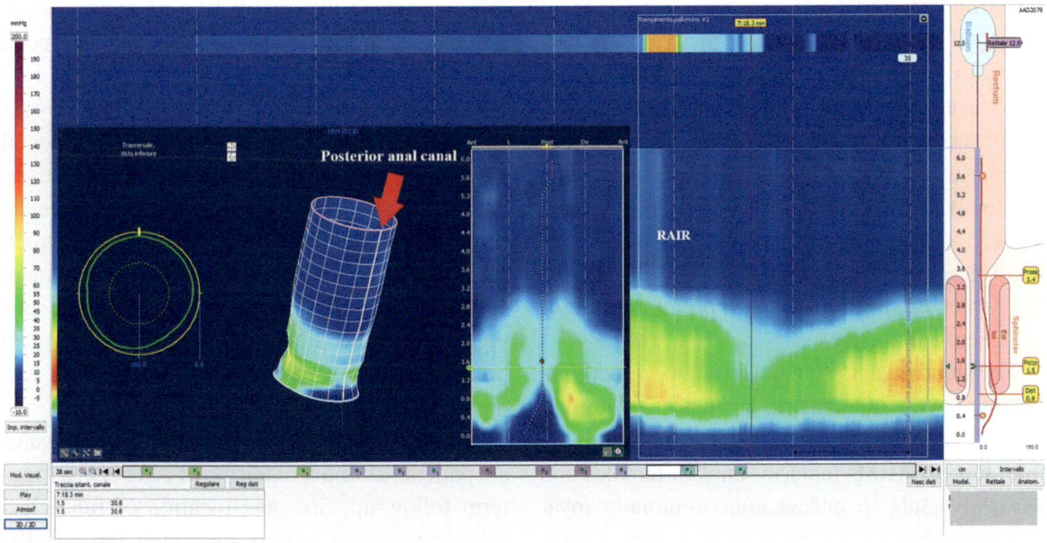

Fig. 8.2 (continued)

sphincter function. The improvement in terms of high-pressure zones in a long-term follow-up was considered secondary to the contractions of the voluntary muscles as intellectual development progresses or as a result of physiotherapeutic training [41].

In contrast to the above, Rintala et al. [42] noted no or very little improvement in sphincter function in terms of decreasing quantity of soiling with increasing age of the patients, probably due to the shorter follow-up. Other manometric and clinical factors were introduced in the evaluation of bowel function: integrity of IAS and presence of sacral and spine congenital anomalies.

Manometrically, poor functional results were associated to low values of anal resting pressure and absence of RAIR which correlate with injuries of IAS. Clinically, patients affected by severe sacral or intraspinal deformities had an unsatisfactory continence outcome, often not related to the concomitant presence of poor sphincter function.

In a comparative manometric study between patients with ARMs and healthy children of different age, lower resting and squeezing pressures with impaired RAIR were demonstrated in all patients of the first group; the length of the anal canal, that contributes to continence, was significantly shorter in operated children who were more than 1 year old [43].

Further studies have underlined the role of anorectal manometry as an objective test compared to the subjective nature of the scoring systems used to evaluate faecal continence in different series of patients affected by ARMs. Based on their own results, clinical continence has been positively correlated with ARP [12, 26, 27, 42, 44, 45], ASP [46], normal rectal sensation [26, 42, 47] and presence of RAIR [26, 42, 44, 48–51], which is considered a good prognostic factor for faecal continence.

Kyrklund et al. [52] reported a good functional outcome with manometric evidence of normal anal pressures and positive RAIR in children with mild type of ARMs (anterior anus in females and perineal fistula in males) after minimally invasive procedures (anal dilatation or conservative follow-up for females with anterior anus and cut-back anoplasty for males with perineal fistula).

Results were poorer among patients with more severe ARMs (vestibular and perineal fistula in females, recto-urethral fistula in males), who had lower ARP and ASP at the manometric test. However, RAIR was found in 83% of these patients after IAS-saving bowel mobilisation surgery (PSARP or anterior sagittal anorectoplasty—ASARP), entailing preservation of the distal fistulous bowel termination. These conclusions confirmed that functional IAS tissue may be found in the termination and that its preservation may influence the continence outcomes, as other researchers had previously detected [53, 54].

Data appear to be confirmed by the absence of RAIR after sacro-perineal and sacro-abdominoperineal interventions which involved resection of the terminal fistulous connection [55].

Additional manometric information descends from profilometric evaluation of the anal canal, defined by a computerised analysis of the pressure curves which is obtained using continuous-flow anorectal manometry. The computer programme generates three-dimensional tracings of the anorectal canal, in order to study the total and segmental asymmetry indices, the rectal volume and also the pressure distribution on the anorectal wall. This test was considered capable of providing reliable information concerning the three-dimensional topography of the anorectal canal and even the distribution of the pressures involved in the process of acquiring anorectal faecal continence.

According to Pedro et al., the first results of this technique showed a total and segmental asymmetry index similar between the continent, partially continent and incontinent patients, proving that surgery was technically correct; greater rectal volumes in partially continent subjects, compatible with constipation and soiling (pseudo-incontinence condition); a typical profilometric pattern with predominance of low pressures in the incontinent group [56].

Recently, 3D-HDARM has been used to evaluate patients after the repairing of ARMs in a long-term follow-up. Size and location of functional sphincter defects were determined through the 3D analysis of the anal resting pressure profile. A functional anal sphincter lesion was defined as a pressure area below 25 mmHg and the size was calculated manually through the 3D cylindrical pressure visualisation. About 50% of the participants showed sphincter defects that in 30% of cases affected more than half of the circumference. Soiling was found only in subjects with sphincter lesions.

A statistically significant correlation was documented between the type of ARMs, size of sphincter defect, mean ASP and Wexner incontinence score.

However, if 3D-HDARM is able to correlate sphincter defects to functional outcome, it is not able to make an anatomic distinction between the different elements of the sphincter complex [57].

8.6 Endoanal Ultrasonography

EAUS is currently considered the gold standard for the morphological evaluation of the anal canal in case of faecal incontinence. Most studies revealed a 100% sensitivity of EAUS in identifying sphincter defects, such as discontinuity, localised or generalised scarring, thinning and atrophy. An endosonographic sphincter lesion is described as an interruption in the normal texture of the muscle ring, while scarring is characterised by loss of texture that usually has low reflectiveness (Figs. 8.3 and 8.4).

Atrophic or degenerative sphincters are seen as thin and poorly defined, with heterogeneous increased echogenicity [58, 59].

On 2D images, the different echogenicity and the numerous interfaces of the structures form-

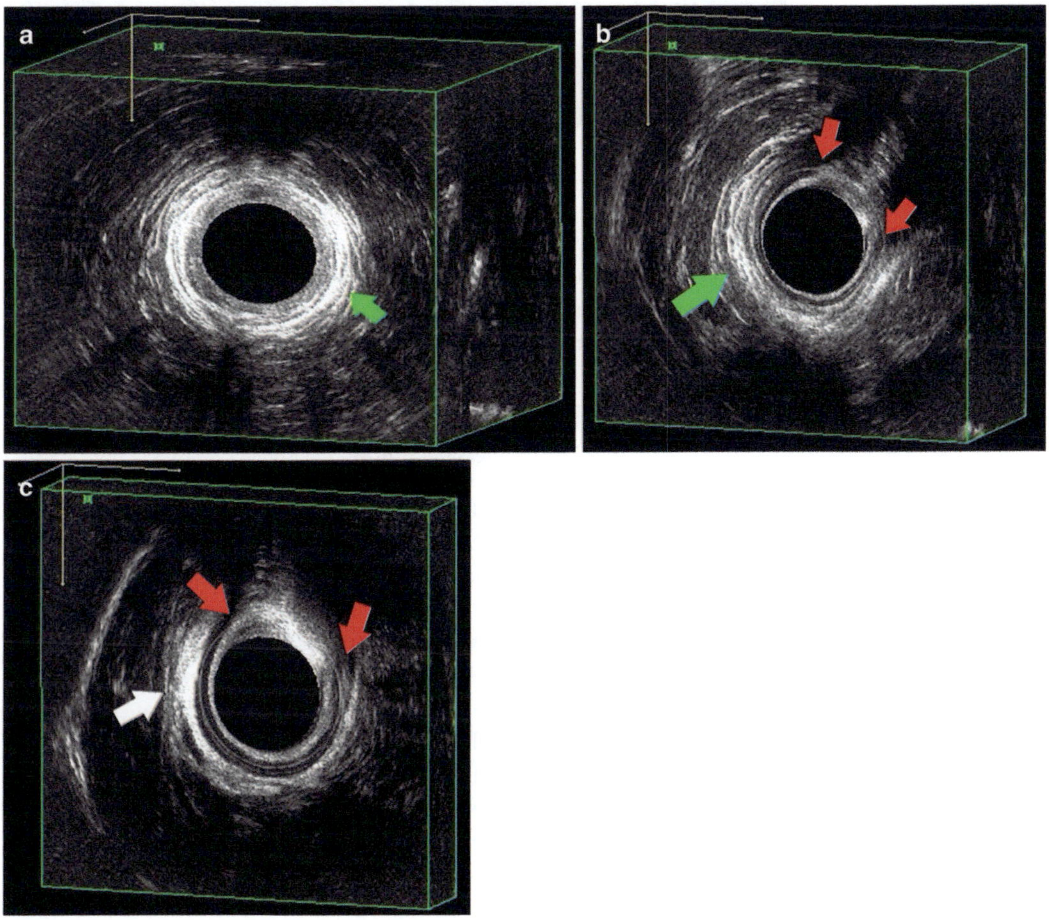

Fig. 8.3 8-year-old male patient with constipation and faecal soiling, surgically treated for anorectal malformation with recto-bulbar fistula (PSARP procedure)
3D-EAUS reveals:
(a) Distal anal canal: generalized scarring of EAS (green arrow)
(b) Middle anal canal: disruption of IAS (red arrows) and generalized scarring of EAS (green arrow)
(c) Proximal anal canal: defect of IAS (red arrows) and generalized scarring of pubo-rectalis muscle (white arrow)
3D-EAUS Three dimensional endoanal ultrasonography, *EAS* External Anal Sphincter, *IAS* Internal Anal Sphincter, *PSARP* Posterior Sagittal AnoRectoPlasty
Source: Digestive Surgery and Endoscopy Unit; Bambino Gesù Children's Hospital-IRCCS-Rome, Italy

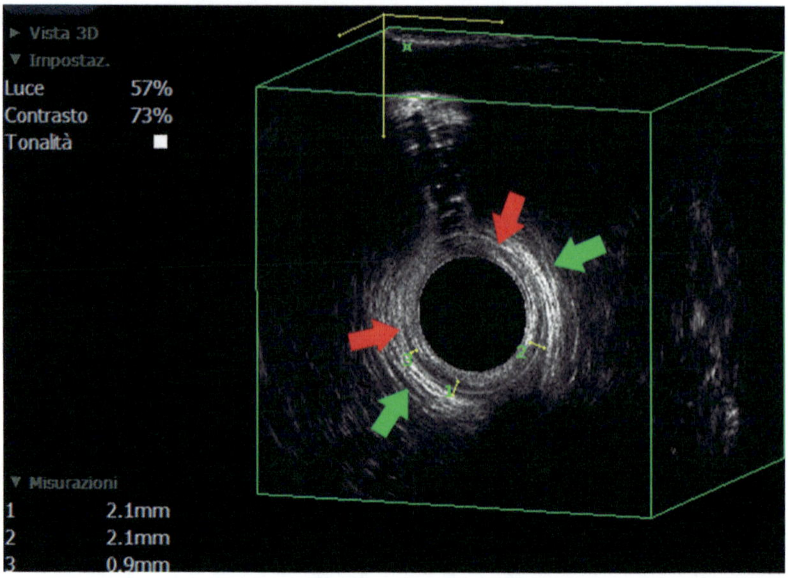

Fig. 8.4 9-year-old female patient with VACTER syndrome and recto-vaginal fistula 3D EAUS shows scars of EAS (green arrows) and lesion of IAS (red arrows); the latter appears thin and disomogeneous in the middle anal canal *3D-EAUS* Three dimensional endoanal ultrasonography, *EAS* External Anal Sphincter, *IAS* Internal Anal Sphincter *Source:* Digestive Surgery and Endoscopy Unit; Bambino Gesù Children's Hospital-IRCCS-Rome, Italy

ing the anal canal account for faithful anatomical depiction of the region and the ability to recognise the single muscular layers composing it [60].

The endoanal transducer consists of a crystal that rotates (4–6 cycles per second) in order to obtain a 360° image. A 6.0-cm-long image is captured along the proximal-distal axis for up to 55 s by moving the crystals on the extremity of the transducer [61].

High-resolution 2D images are usually recorded at three different levels of the anal canal (proximal, middle and distal), typified by the presence of [62–64]:

- IAS and puborectalis muscle appearing, respectively, as a hypoechoic ring and a hyperechoic horseshoe sling in the proximal anal canal.
- IAS and EAS in the middle anal canal: EAS forms a broad and mixed echogenicity ring tending to hyperechogenicity, which lies immediately outside the IAS.
- EAS in the distal anal canal.

The advent of 3D technology has further improved the understanding of the two-dimensional technique, also solving the downside of being an operator-dependent study.

3D reconstruction provides a multiplanar imaging of the anal canal, allowing length, thickness, area and volume measurement. After a 3D dataset has been acquired, it is possible to select coronal anterior-posterior or posterior-anterior as well as sagittal right–left views, together with any oblique image plane. The 3D image, showed as a "cube", can be rotated, tilted and sliced in any other direction to enable visualisation from different angles. This yields more information on the anal sphincter complex and makes it easier to perform sphincter measurements [63–66].

Several studies have compared EAUS with MRI, concluding that 3D-EAUS is superior in diagnosing IAS injury, equivalent in detecting EAS injury and inferior in identifying EAS atrophy [67].

A paediatric study on patients surgically treated for ARMs confirmed that EAUS has higher accuracy than MRI in recognising slight malposition of the neo-anus within the striated muscle complex and sphincter defects in subjects with abnormal muscle contraction on perineal muscle stimulation [68].

Besides, 3D EAUS has the plus of being easier, quicker, cheaper and better tolerated by patients than MRI; therefore it may be considered the method of choice [69] even in children.

Although a medical technician may also be able to technically perform EAUS, it is recommended that the procedure is carried out by a specialist to ensure the best insight of the awaited anatomy in accord to the underlying disease.

The joint committee of the Italian Society of Colorectal Surgery (SICCR) and the Italian Society of Ultrasonology in Medicine and Biology (SIUMB) defined the endoanal sonographic centres' accreditation requirements and established that physicians can be considered "expert" if they usually perform at least 5–6 studies per week [70].

An accurate inspection of the perineum and a digital rectal examination should be conducted to exclude severe stenosis of the anal opening and to avoid complications before 3D-EAUS, especially after reconstructive surgery of the anus.

Sedation is not indispensable in adulthood, but may be necessary for younger or complaining children.

8.7 Endoanal Ultrasonography and Anorectal Manometry in Anorectal Malformations

EAUS and anorectal manometry are combined investigations that should be used together to perform a complete estimation of anal sphincters in case of evacuatory dysfunctions (Figs. 8.5 and 8.6).

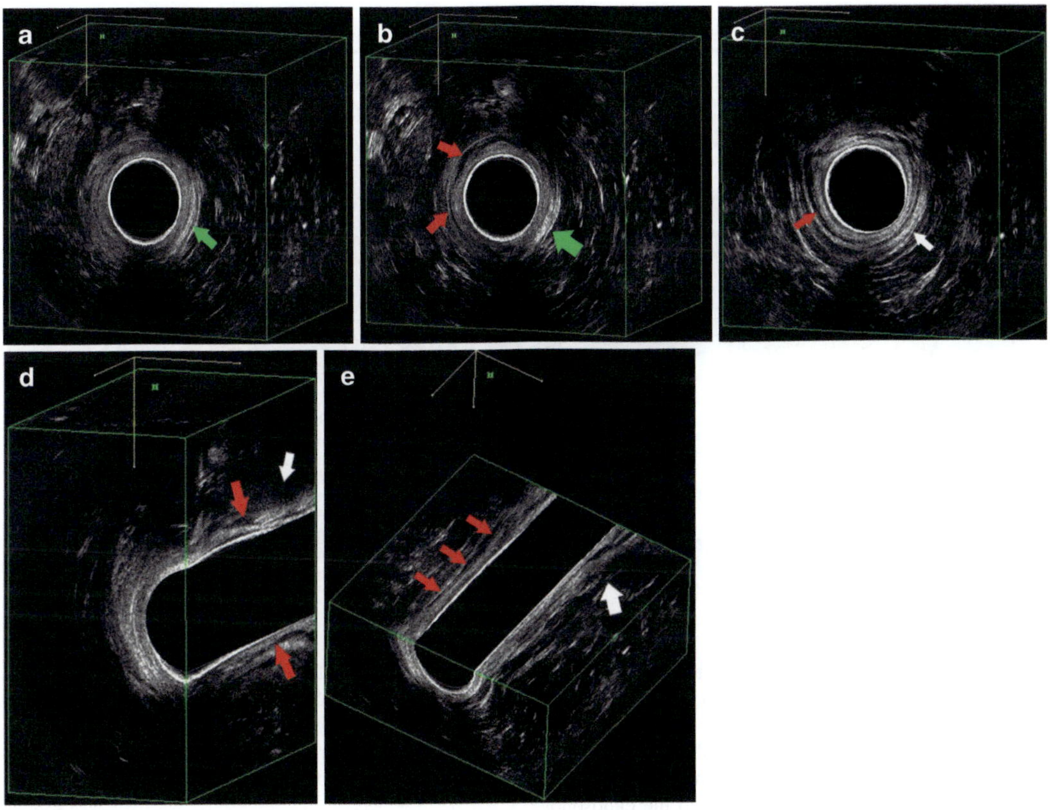

Fig. 8.5 3D endosonographic pattern of a 17-year-old female patient with VACTER syndrome and faecal incontinence
Axial sections:
- (**a**) Distal anal canal: localized scarring of EAS (green arrow)
- (**b**) Middle anal canal: fragmented IAS with only a few remnants (red arrows); scarring of EAS (green arrow)
- (**c**) Proximal anal canal: scar tissue (white arrow) of puborectalis muscle; fragmentation of IAS (red arrow)

Sagittal (**d**) and coronal (**e**) planes: thin fragments of IAS (red arrows) and atrophy of puborectalis muscle (white arrow)
EAS External Anal Sphincter; *IAS* Internal Anal Sphincter
Source: Digestive Surgery and Endoscopy Unit; Bambino Gesù Children's Hospital-IRCCS-Rome, Italy

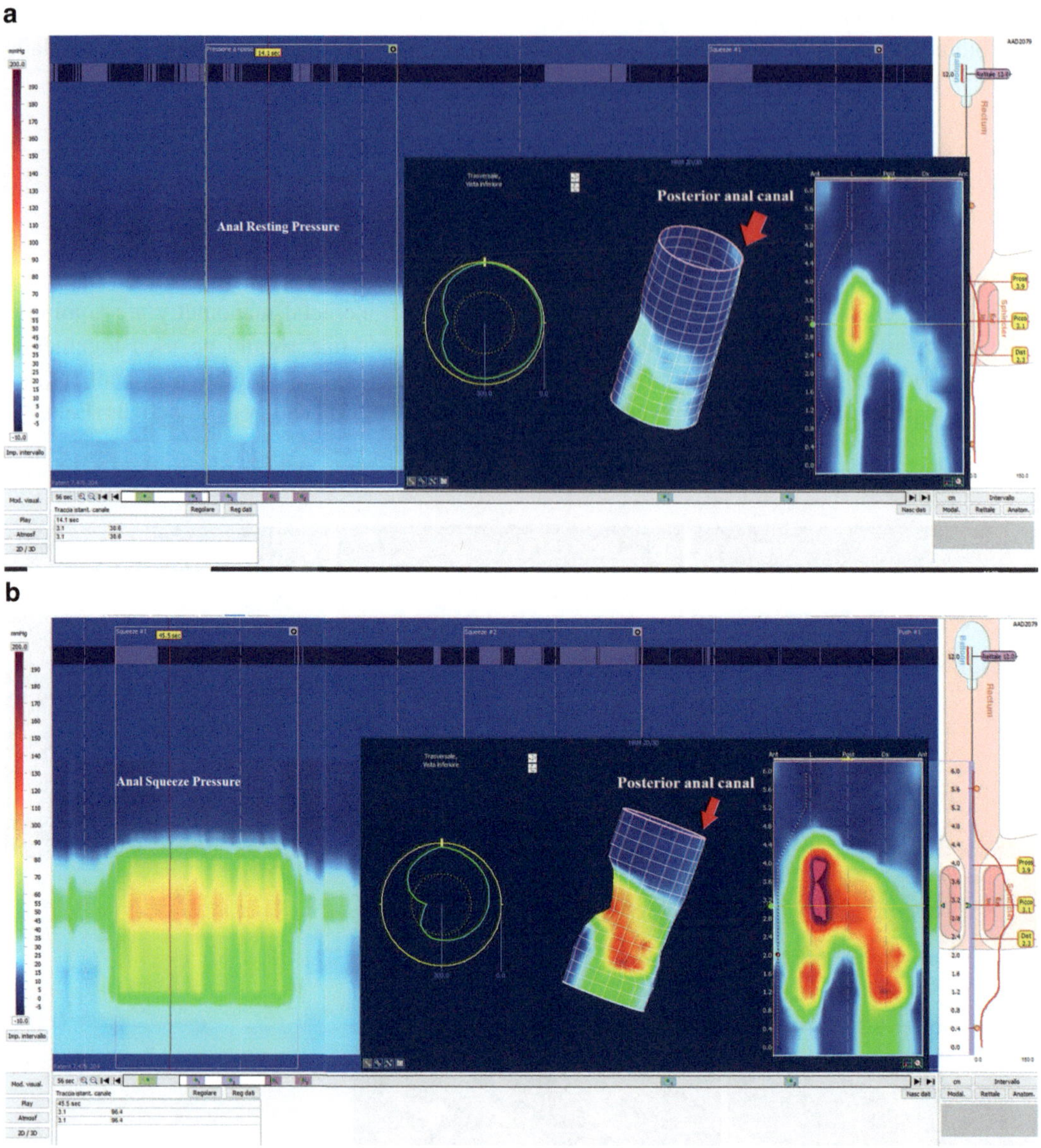

Fig. 8.6 High definition manometric pattern of a 17-year-old female patient with VACTER syndrome and faecal incontinence
HD-ARM shows low ARP (**a**) and ASP (**b**), no signs of pelvic floor dyssynergia (**c**) and absence of RAIR (**d**). Data of HD-ARM are complementary to the endosonographic findings (Fig. 8.5)
HD-ARM High definition anorectal manometry, *ARP* Anal Resting Pressure, *ASP* Anal Squeeze Pressure, *RAIR* Recto-Anal Inhibitory Reflex
Source: Digestive Surgery and Endoscopy Unit; Bambino Gesù Children's Hospital-IRCCS-Rome, Italy

c

d

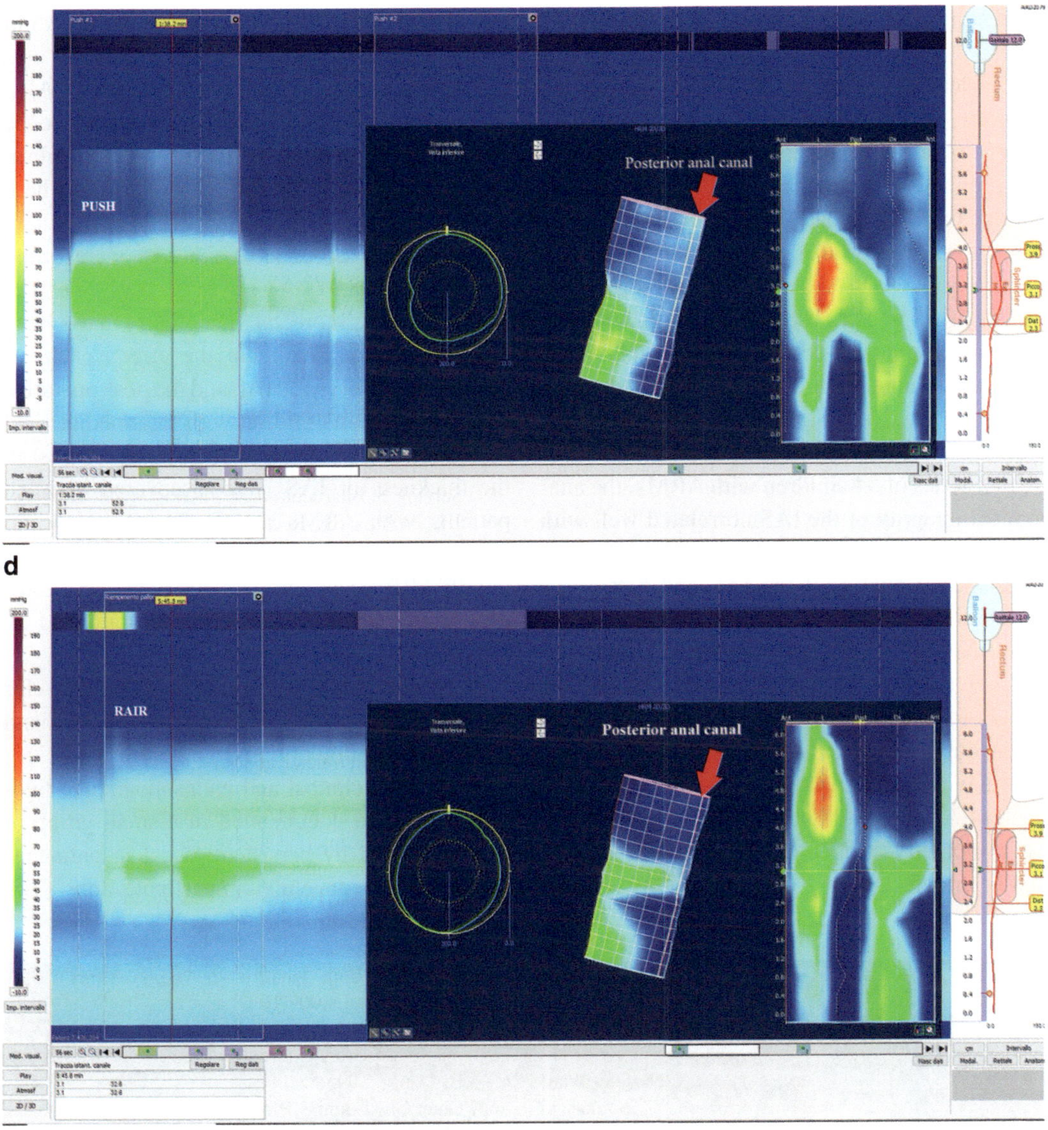

Fig. 8.6 (continued)

Although faecal incontinence may be the result of several causes, anal sphincter injury is highly prevalent, particularly after anorectal surgery.

The two procedures offer an objective baseline assessment of anal sphincter, and the results obtained may help to define the therapeutic medical or surgical programme for the improvement of faecal incontinence and constipation.

Until now, only a few studies have evaluated patients who had undergone reconstructive surgery for anorectal anomalies by EAUS, but very interesting data have been derived from these series.

Emblem et al. found a strong correlation between clinical, manometric and endosonographic findings in patients with ARMs and a control group [71]. Specifically, children with high/intermediate ARMs had a poorer continence outcome, lower ARP and ASP and sphincter defects consisting of scars of IAS and EAS, some remnants of EAS and/or absence of IAS. In addition, the sphincter muscle complex and its relation to the anal opening were visualised by anal endosonography and varying degrees of eccentrically placed anal canal in the EAS were identified.

In a series of 54 children with ARMs, the anatomical integrity of the IAS correlated well with ARP, RAIR and type of ARMs (low-ARM group had superior quality of the IAS and better faecal continence than the one with high ARMs). Nevertheless, megarectum and and/or neuropathy are confirmed as unfavourable prognostic factors for faecal dynamics because their occurrence outweighs the benefit of good IAS and causes incontinence [48].

3D-EAUS could be superior to anorectal manometry in identifying mild and modest sphincter defects (Fig. 8.7).

As Caldaro et al. [12] reported, 3D-EAUS showed small/moderate disruptions of the IAS in a high percentage of cases with low ARMs, notwithstanding normal ARP and presence of RAIR.

In severe types of ARMs, correspondence between symptoms, manometric data and ultrasound findings was more significant; in these subjects, generalised scarring of the EAS, width defects and/or absence of the IAS were found (Figs. 8.8 and 8.9).

EAUS and anorectal manometry have been utilised even to document the effectiveness of PSARP to preserve the IAS and the post-operative anal functions in children with intermediate and high defects. In fact, if major differences in the thickness of IAS were discovered between patients with ARMs and healthy controls, no substantial dissimilarities were found between the PSARP group and transperineal anorectoplasty one, which components were affected by low lesions [47].

Both procedures are also useful exams for monitoring patients in the follow-up period and to plan therapeutic options.

Interesting clinical and endosonographic findings have been documented in a small group of adults affected by persistent faecal incontinence secondary to congenital imperforate anus [72].

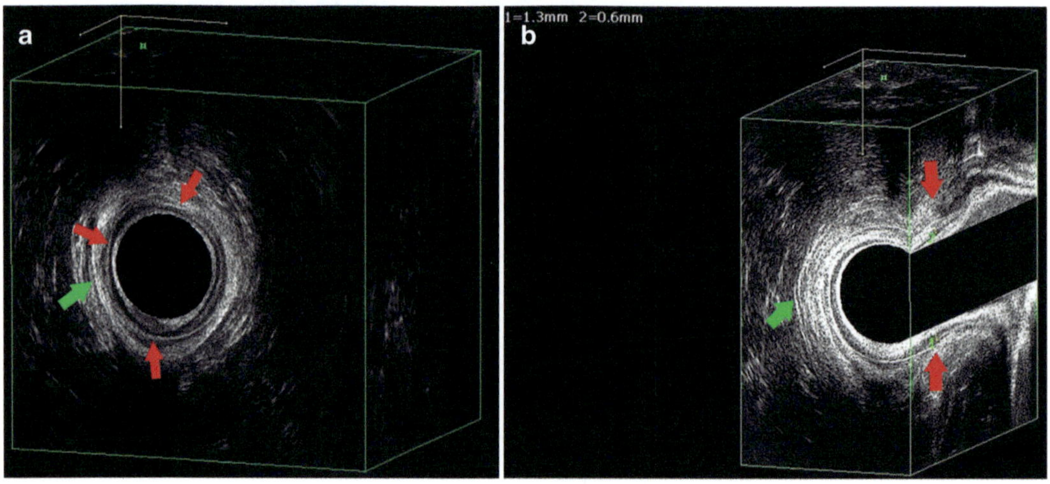

Fig. 8.7 3D EAUS in 5-year-old female patient operated on vestibular fistula, with chronic constipation
Axial (**a**) and sagittal (**b**) planes: Scarring of EAS (green arrow) and irregular thickness of IAS (red arrows)
3D-EAUS Three dimensional endoanal ultrasonography, *EAS* External Anal Sphincter, *IAS* Internal Anal Sphincter
Source: Digestive Surgery and Endoscopy Unit; Bambino Gesù Children's Hospital-IRCCS-Rome, Italy

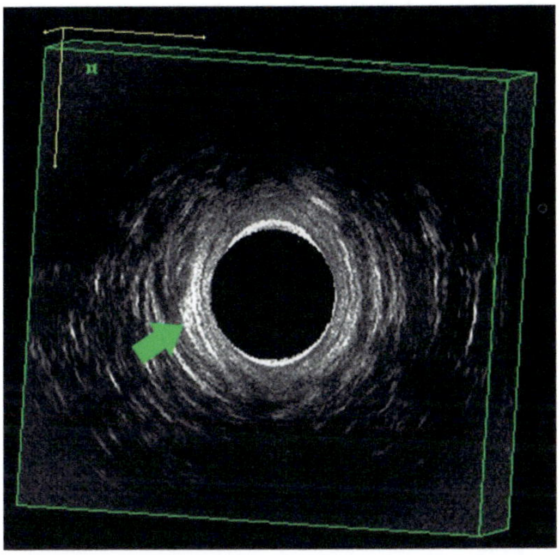

Fig. 8.8 9-year-old male patient with anorectal malformations and recto-bladder neck fistula. Clinical history of severe faecal incontinence
EAUS detects abundant scar tissue of the EAS (green arrow) and absence of the IAS in middle anal canal
EAUS Endoanal ultrasonography, *EAS* External Anal Sphincter, *IAS* Internal Anal Sphincter
Source: Digestive Surgery and Endoscopy Unit; Bambino Gesù Children's Hospital-IRCCS-Rome, Italy

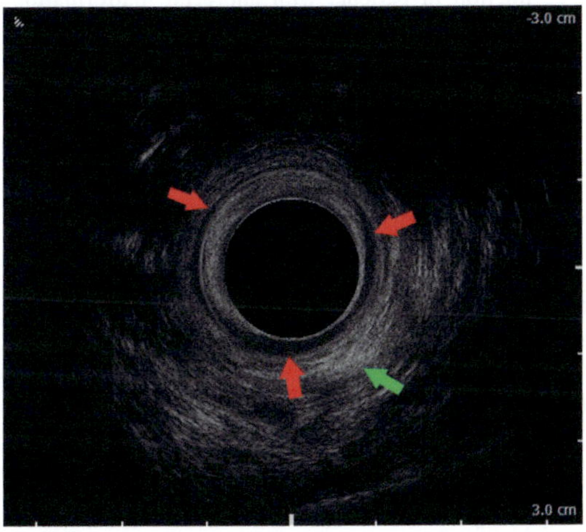

Fig. 8.9 17-year-old male patient with faecal incontinence secondary to anorectal malformation and recto-bulbar fistula
2D image of middle anal canal: EAUS identifies irregular thickness of the IAS (red arrows) and scarring of the EAS (green arrow)
EAUS Endoanal ultrasonography, *EAS* External Anal Sphincter, *IAS* Internal Anal Sphincter
Source: Digestive Surgery and Endoscopy Unit; Bambino Gesù Children's Hospital-IRCCS-Rome, Italy

Abnormal clinical evidences (e.g. anal stricture, prolapse, misplaced neo-anus) and EAS injuries were found in more than 90% and in 50% of cases, respectively. Conservative or surgical therapeutic choices have been proposed according to an algorithm of recommended management in relation to the causes of incontinence. Therefore, medical treatment was administered or optimised when absent or incongruous; biofeedback training was suggested when a patient with an intact sphincter was unable to contract it on demand and surgery was indicated in case of prolapse or anal stricture. Other minimally invasive therapies, as injection of bulking agents or sacral nerve stimulation (SNS), were advocated in non-responder subjects.

Injectable bulking therapy (non-animal stabilised hyaluronic acid with dextranomer-NASHA/Dx) has been used with a significant effect on the number of incontinence episodes in adults with rectobulbar or vestibular fistula [73]. All patients were assessed preoperatively and in follow-up by 3D-EAUS to evaluate migration of the implants.

SNS seems to be another promising alternative treatment [74]. Data are partial and heterogeneous, but encouraging results have been achieved in subjects with intact IAS, low ARP and ASP and even in the presence of partial sacral agenesis, which could on the other hand complicate the placement of stimulator leads in the foramina of the sacrum [75, 76].

Patients suffering from such severe incontinence that has not been possible to amend with other less invasive measures could be candidates to surgical correction. In fact, although the outcome of dynamic graciloplasty in cases of ARMs was proved inferior compared to the ones with other underlying aetiologies, it may be indicated in selected occurrences in which forming a stoma is the next step [77].

8.8 Conclusions

Faecal incontinence and chronic constipation are disabling conditions due to congenital, anatomical and post-surgical factors in patients operated on ARMs, with impact on the quality of life [78].

In adulthood, anorectal manometry and EAUS are validated tests in the management of anorectal disorders. EAUS has a high degree of sensitivity and specificity and correlates well with manometric findings, so both investigations are considered complementary in assessing the morphology and function of anal sphincters [79].

In paediatrics, instead, controlled prospective studies aiming to evaluate the impact of these exams on treatments and long-term outcomes are still limited.

However, the results achieved so far have provided useful information to define the anorectal pathophysiology, even in patients with AMRs; consequently it should be mandatory to consider EAUS and manometry in the diagnostic algorithm of evacuatory dysfunction.

References

1. Rigueros Springford L, Connor MJ, Jones K, Kapetanakis VV, Giuliani S. Prevalence of active long-term problems in patients with anorectal malformations: a systematic review. Dis Colon Rectum. 2016;59:570–80. https://doi.org/10.1097/DCR.0000000000000576.
2. Ludman L, Spitz L, Kiely EM. Social and emotional impact of fecal incontinence after surgery for anorectal abnormalities. Arch Dis Child. 1994;71:194–200.
3. Minneci PC, Kabre RS, Mak GZ, Halleran DR, Cooper JN, Afrazi A, et al. Can fecal continence be predicted in patients born with anorectal malformations? J Pediatr Surg. 2019;54:1159–63. https://doi.org/10.1016/j.jpedsurg.2019.02.035.
4. Wood RJ, Marc A, Levitt MA. Anorectal malformations. Clin Colon Rectal Surg. 2018;31:61–70.
5. Tantiphlachiva K. Comprehensive clinical approach to fecal incontinence. Book chapter doi: https://doi.org/10.5772/intechopen.86346.
6. Frischer JS, Rymeski B. Complications in colorectal surgery. Semin Pediatr Surg. 2016;25:380–7. https://doi.org/10.1053/j.sempedsurg.2016.10.008.
7. Stephens FD, Smith ED. Classification, identification, and assessment of surgical treatment of anorectal anomalies. Pediatr Surg Int. 1986;1:200–5.
8. Peña A. Anorectal malformations. Semin Pediatr Surg. 1995;4:35–47.
9. Holschneider A, Hutson J, Peña A, Beket E, Chatterjee S, Coran A, et al. Preliminary report on the international conference for the development of standards for the treatment of anorectal malformations. J Pediatr Surg. 2005;40:1521–6. https://doi.org/10.1016/j.jpedsurg.2005.08.002.
10. Gupta DK. Anorectal malformations—wingspread to Krickenbeck. J Indian Assoc of Pediatr Surg. 2005;10:75–7.
11. Guillaume A, Salem AE, Garcia P, Chander RB. Pathophysiology and therapeutic options for fecal incontinence. J Clin Gastroenterol. 2017;51:324–30. https://doi.org/10.1097/MCG.0000000000000797.
12. Caldaro T, Romeo E, De Angelis P, Gambitta RA, Rea F, Torroni F, et al. Three-dimensional endoanal ultrasound and anorectal manometry in children with anorectal malformations: new discoveries. J Pediatr Surg. 2012;47:956–63.
13. Borg H, Bachelard M, Sillén U. Megarectosigmoid in children with anorectal malformations: long term outcome after surgical or conservative treatment. J Pediatr Surg. 2014;49:564–9. https://doi.org/10.1016/j.jpedsurg.2013.08.003.
14. Burjonrappa S, Youssef S, Lapierre S, Bensoussan A, Bouchard S. Megarectum after surgery for anorectal malformations. J Pediatr Surg. 2010;45:762–8. https://doi.org/10.1016/j.jpedsurg.2009.10.043.
15. Miyahara K, Kato Y, Seki T, Arakawa A, Lane GJ, Yamataka A. Neuronal immaturity in normoganglionic colon from cases of Hirschsprung disease, anorectal malformation, and idiopathic constipation. J Pediatr Surg. 2009;44:2364–8. https://doi.org/10.1016/j.jpedsurg.2009.07.066.
16. Holbrook C, Misra D, Zaparackaite I, Cleeve S. Postoperative strictures in anorectal malformation: trends over 15 years. Pediatr Surg Int. 2017;33:869–73. https://doi.org/10.1007/s00383-017-4111-6.
17. Kyrklund K, Pakarinen MP, Rintala RJ. Long-term bowel function, quality of life and sexual function in patients with anorectal malformations treated during the PSARP era. Semin Pediatr Surg. 2017;26:336–42. https://doi.org/10.1053/j.sempedsurg.2017.09.010.
18. Peña A, Bischoff A. Surgical treatment of colorectal problems in children. Springer; 2015. https://doi.org/10.1007/978-3-319-14989-9.
19. Gurusamy R, Raj SV, Maniam R, Regunandan SR. Laparoscopic-assisted anorectoplasty: a single-center experience. J Indian Assoc Pediatr Surg. 2017;22:114–8.
20. Peña A, Hong A. Advances in the management of anorectal malformations. Am J Surg. 2000;180:370–6. https://doi.org/10.1016/s0002-9610(00)00491-8.
21. Pathak M, Saxena AK. Postoperative "complications" following laparoscopic-assisted anorectoplasty: a systematic review. Pediatr Surg Int. 2020;36:1299–307. https://doi.org/10.1007/s00383-020-04748-3.
22. Rodriguez L, Sood M, Di Lorenzo C, Saps M. An ANMS-NASPGHAN consensus document on anorectal and colonic manometry in children. Neurogastroenterol Motil. 2017;29:10.1111.
23. Rasijeff AMP, Withers M, Burke JM, Jackson W, S. Scott SM. High-resolution anorectal manometry: a comparison of solid-state and water-perfused catheters. Neurogastroenterol Motil. 2017;29:e13124.
24. Scott SM, Emma V, Carrington EV. The London classification: improving characterization and classification of anorectal function with anorectal manometry. Curr Gastroenterol Rep. 2020;22:–55. https://doi.org/10.1007/s11894-020-00793-z.

25. Tambucci R, Quitadamo P, Thapar N, Zenzeri L, Caldaro T, Staiano A, Verrotti A, Borrelli O. Diagnostic tests in pediatric constipation. J Pediatr Gastroenterol Nutr. 2018;66(4):e89–98. https://doi.org/10.1097/MPG.0000000000001874.
26. Hedlund H, Peña A, Rodriguez G, Maza J. Long-term anorectal function in imperforate anus treated by a posterior sagittal anorectoplasty: manometric investigation. J Pediatr Surg. 1992;27:906–9.
27. Senel E, Demirbag S, Tiryaki T, Erdogan D, Cetinkursun S, Cakmak O. Postoperative anorectal manometric evaluation of patients with anorectal malformation. Pediatr Int. 2007;49:210–4. https://doi.org/10.1111/j.1442-200X.2007.02342.x.
28. Rintala RJ. Fecal incontinence in anorectal malformations, neuropathy, and miscellaneous conditions. Semin Pediatr Surg. 2002;11:75–82.
29. Lee TH, Bharucha AE. How to perform and interpret a high-resolution anorectal manometry test. J Neurogastroenterol Motil. 2016;31(22):46–59. https://doi.org/10.5056/jnm15168.
30. William PT, John CA, Patrick DB. Neuroradiologic evaluation of sacral abnormalities in imperforate anus complex. J Pediatr Surg. 1987;22:58–61.
31. Capitanucci ML, Rivosecchi M, Silveri M, Lucchetti MC, Mosiello G, De Gennaro M. Neurovesical dysfunction due to spinal dysraphism in anorectal anomalies. Eur J Pediatr Surg. 1996;6:159–62.
32. Yuan Z, Bai Y, Zhang Z, Ji S, Li Z, Wang W. Neural electrophysiological studies on the external anal sphincter in children with anorectal malformation. J Pediatr Surg. 2000;35:1052–7. https://doi.org/10.1053/jpsu.2000.7770.
33. Chiarioni G, Kim SM, Vantini I, Whitehead WE. Validation of the balloon evacuation test: reproducibility and agreement with findings from anorectal manometry and electromyography. Clin Gastroenterol Hepatol. 2014;12:2049–54.
34. Rao SS. Dyssynergic defecation and biofeedback therapy. Gastroenterol Clin N Am. 2008;37:569–86. viii. https://doi.org/10.1016/j.gtc.2008.06.011.
35. Rao SS, Ozturk R, Laine L. Clinical utility of diagnostic tests for constipation in adults: a systematic review. Am J Gastroenterol. 2005;100:1605–15.
36. Palsson OS, Heymen S, Whitehead WE. Biofeedback treatment for functional anorectal disorders: a comprehensive efficacy review. Appl Psychophysiol Biofeedback. 2004;29:153–74.
37. Rao SS, Tuteja AK, Vellema T, et al. Dyssynergic defecation: demographics, symptoms, stool patterns, and quality of life. J Clin Gastroenterol. 2004;38:680–5.
38. van Meegdenburg MM, Heineman E, Broens PMA. Dyssynergic defecation may aggravate constipation: results of mostly pediatric cases with congenital anorectal malformation. Am J Surg. 2015;210:357–64.
39. Bjørsum-Meyer T, Christensen P, Baatrup G, Jakobsen MS, Asmussen J, QvistN. Dyssynergic patterns of defecation in constipated adolescents and young adults with anorectal malformations. Sci Rep. 2020;10(19673) https://doi.org/10.1038/s41598-020-76841-5.
40. Iwai N, Ogita S, Kida M, Fujita Y, Majima S. A clinical and manometric correlation for assessment of postoperative continence in imperforate anus. J Pediatr Surg. 1979;14:538–43.
41. Iwai N, Yanagihara J, Tokiwa K, Deguchi E, Takahashi T. Results of surgical correction of anorectal malformations. A 10-30 year follow-up. Ann Surg. 1988;207:219–22.
42. Rintala RJ, Lindahl H. Is Normal bowel function possible after repair of intermediate and high anorectal malformations? J Pediatr Surg. 1995;30:491–4.
43. Kumar S, Ramadan SA, Gupta V, Helmy S, Debnath P, Alkholy A. Use of anorectal manometry for evaluation of postoperative results of patients with anorectal malformation: a study from Kuwait. J Pediatr Surg. 2010;45:1843–8.
44. Iwai N, Hashimoto K, Goto Y, Majima S. Long-term results after surgical correction of anorectal malformations. Z Kinderchir. 1984;39:35–9.
45. Heikenen JB, Werlin SL, Di Lorenzo C, Hyman PE, Cocjin J, Flores AF, et al. Colonic motility in children with repaired imperforate anus. Dig Dis Sci. 1999;44:1288–92.
46. Molander ML, Frenckner B. Anal sphincter function after surgery for high imperforate anus—a long term follow-up investigation. Z Kinderchir. 1985;40:91–6.
47. Wang Z, Hu L, Jin X, Li X, Xu L. Evaluation of postoperative anal functions using endoanal ultrasonography and anorectal manometry in children with congenital anorectal malformations. J Pediatr Surg. 2016;51:416–20.
48. Keshtgar AS, Athanasakos E, Clayden GS, Ward HC. Evaluation of outcome of anorectal anomaly in childhood: the role of anorectal manometry and endosonography. Pediatr Surg Int. 2008;24:885–92.
49. Aspirot A. Anorectal malformations. In: Faure C, Thapar N, Di Lorenzo C, editors. Pediatric neurogastroenterology. Cham: Springer; 2017. https://doi.org/10.1007/978-3-319-43268-7_29.
50. Kelly JH. The clinical and radiological assessment of anal continence in childhood. Aust NZJ Surg. 1972;42:62–3.
51. Kiesewetter WB, Chang JH. Imperforate anus: a five to thirty-year follow-up perspective. Prog Pediatr Surg. 1977;10:111–20.
52. Kyrklund K, Pakarinen MP, Rintala RJ. Manometric findings in relation to functional outcomes in different types of anorectal malformations. J Pediatr Surg. 2017;52:563–8.
53. Kluth D, Hillen M, Lamprecht W. The principles of normal and abnormal hindgut development. J Pediatr Surg. 1995;30:1143–7.
54. Ruttenstock EM, Zani A, Huber-Zeyringer A, Höllwarth ME. Pre- and postoperative rectal manometric assessment of patients with anorectal malformations: should we preserve the fistula? Dis Colon Rectum. 2013;56:499–504. https://doi.org/10.1097/DCR.0b013e31826e4a38.

55. Rintala RJ, Lindahl HG. Posterior sagittal anorectoplasty is superior to sacroperineal sacroabdominoperineal pull-through: a long-term follow-up study in boys with high anorectal anomalies. J Pediatr Surg. 1999;34:334–7.
56. Júnior PFV, Martins JL, Peterlini FL. Posterior sagittal anorectoplasty in anorectal anomalies: clinical, manometric and profilometric evaluation. Sao Paulo Med J. 2007;125:163–9.
57. Bjørsum-Meyer T, Christensen P, Jakobsen MS, Baatrup G, Qvist N. Correlation of anorectal manometry measures to severity of fecal incontinence in patients with anorectal malformations—a cross sectional study. Sci Rep. 2020;10:19673. https://doi.org/10.1038/s41598-020-76841-5.
58. Sultan AH, Kamm MA, Talbot IC, Nicholls RJ, Bartram CI. Anal endosonography for identifying external sphincter defects confirmed histologically. Br J Surg. 1994;81:463–5.
59. Nichols CM, Lamb EH, Ramakrishnan V. Differences in outcomes after third-versus fourth-degree perineal laceration repair: a prospective study. Am J Obstet Gynecol. 2005;193:530–4.
60. Reginelli A, Mandato Y, Cavaliere C, Pizza NL, Russo A, Cappabianca S, et al. Three-dimensional anal endosonography in depicting anal-canal anatomy. Radiol Med. 2012;117:759–71.
61. Murad Regadas SM, Regadas FS, Rodrigues LV, Holanda EC, Barreto RG, et al. The role of 3-dimensional anorectal ultrasonography in the assessment of anterior trans-sphincteric fistula. Dis Colon Rectum. 2010;53:1035–40.
62. Hussain SM, Stoker J, Laméris JS. Anal sphincter complex: endoanal MR imaging of normal anatomy. Radiology. 1995;197:671–7.
63. Abdool Z, Sultan AH, Thakar R. Ultrasound imaging of the anal sphincter complex: a review. Br J Radiol. 2012;85:865–75.
64. Villanueva Herrero JA, Hansen MR, Bernes LA, Bobadill J. Endoanal ultrasound for anorectal diseases. In: Gastrointestinal diseases. SM Group. p. 1–23.
65. Albuquerque A. Endoanal ultrasonography in fecal incontinence: current and future perspectives. World J Gastrointest Endosc. 2015;10:575–81.
66. Gosselink MP, West RL, Kuipers EJ, Hansen BE, Schouten WR. Integrity of the anal sphincters after pouch-anal anastomosis: evaluation with three-dimensional endoanal ultrasonography. Dis Colon Rectum. 2005;48:1728–35.
67. Nuernberg D, Saftoiu A, Barreiros AP, Burmester E, Ivan ET, Clevert DA, et al. EFSUMB recommendations for gastrointestinal ultrasound part 3: endorectal. Endoanal Perineal Ultrasound Int Open. 2019;5:E34–51.
68. Athanasakos EP, Ward HC, Williams NS, Scott SM. The value of anal endosonography compared with magnetic resonance imaging following the repair of anorectal malformations. Pediatr Radiol. 2003;33:183–5.
69. Emblem R, Morkrid L, Bjornland K. Anal endosonography is useful for postoperative assessment of anorectal malformations. J Pediatr Surg. 2007;42:1549–54.
70. Dal Corso HM, D'Elia A, De Nardi P, Cavallari F, Favetta U, et al. Anal endosonography. A survey of equipment, technique and diagnostic criteria adopted in nine Italian centers. Tech Coloproctol. 2007;11:26–33.
71. Emblem R, Diseth T, Morkrid L. Anorectal anomalies: anorectal manometric function and anal endosonography in relation to functional outcome. Pediatr Surg Int. 1997;12:516–9.
72. Danielson J, Karlbom U, Graf W, Wester T. Persistent fecal incontinence into adulthood after repair of anorectal malformations. Int J Color Dis. 2019;34(3):551–4. https://doi.org/10.1007/s00384-018-3220-6.
73. Danielson J, Karlbom U, Wester T, Graf W. Injectable bulking treatment of persistent fecal incontinence in adult patients after anorectal malformations. J Pediatr Surg. 2020;55:397–402.
74. Dewberry L, Trecartin A, Peña A, Pierre MS, Bischoff A. Systematic review: sacral nerve stimulation in the treatment of constipation and fecal incontinence in children with emphasis in anorectal malformation. Pediatr Surg Int. 2019;35:1009–12. https://doi.org/10.1007/s00383-019-04515-z.
75. Thomas GP, Nicholls RJ, Vaizey CJ. Sacral nerve stimulation for faecal incontinence secondary to congenital imperforate anus. Tech Coloproctol. 2013;17:227–9.
76. Eftaiha SM, Melich G, Pai A, Marecik SJ, Prasad LM, Park JJ. Sacral nerve stimulation in the treatment of bowel dysfunction from imperforate anus: a case report. Int J Surg Case Rep. 2016;24:115–8. https://doi.org/10.1016/j.ijscr.2016.05.018.
77. Danielson J, Karlbom U, Wester T, Graf W. Long-term outcome after dynamic graciloplasty for treatment of persistent fecal incontinence in patients with anorectal malformations. Eur J Pediatr Surg. 2019;29(3):276–81. https://doi.org/10.1055/s-0038-1641599.
78. Reddymasu SC, Singh S, Waheed S, Oropeza-Vail M, McCallum RW, Olyaee M. Comparison of anorectal manometry to endoanal ultrasound in the evaluation of fecal incontinence. Am J Med Sci. 2009;337:336–9.
79. Bordeianou L, Lee K, Rockwood T, Baxter N, Lowry A, Mellgren A, et al. Anal resting pressures at manometry correlate with the fecal incontinence severity index and with presence of sphincter defects on ultrasound. Dis Colon Rectum. 2018;51:1010–4.

9 Suppurative Processes of Anorectal Region: Endoanal Ultrasonographic Assistance

Dajana Cuicchi and Gilberto Poggioli

Suppurative processes of the anorectal region are a common condition, ranging from acute abscess to chronic fistula formation. A perianal fistula is formally defined as a chronic track of granulation tissue which connects the anal canal with the perianal skin [1]. The majority of perianal fistulas, up to 90%, are idiopathic fistulas related to cryptoglandular disease; fewer than 10% are fistulas associated with chronic intestinal inflammatory diseases (such as Crohn's) and finally, a small percentage of anal fistulas are the consequence of trauma (including foreign body injury or obstetric trauma), infections (such as tuberculosis) and iatrogenic lesions [2]. According to a slightly adapted version of Parks classification, perianal fistulas are classified into submucosal (15%), intersphincteric (24%), transsphincteric (58%), suprasphincteric (3%) or extrasphincteric (<1%), depending on the muscle striated which surpasses [3, 4].

Since the endoanal ultrasound (EAUS) technique was first described by Bartram et al. in 1989 and was used to obtain high-resolution images of the anal canal, it was immediately widely used for the study of anal fistulas [5, 6]. Ultrasound allows for accurately defining the topography of the fistula tract and its relationship with the anal sphincters, detecting secondary extensions, and furnishing information regarding the existence of anal sphincter defects before surgery. All this information is critical to reducing the risk of septic recurrence and post-operative faecal incontinence. Since this technique has been shown to be superior to clinical examination alone in the diagnosis and classification of anorectal suppurative processes, EAUS is recommended in patients with suspected or proven anal fistulas [7, 8].

At EAUS, the fistula appears as a linear hypoechoic band in the longitudinal sections and as a round hypoechoic area in the axial sections (Fig. 9.1) [9]. It may contain small amounts of gas within it which appear as hyperechoic spots with a posterior shadow [9]. The primary tract of a fistula should be described according to Parks classification, and the internal and external orifice should be located and any extensions identified. The criteria for identifying the internal opening of a fistula in ano include a hypoechoic breach of the subepithelial layer, a defect in the circular smooth muscle of the internal anal sphincter (IAS) and a hypoechoic lesion in the normally hyperechoic intersphincteric space [10]. Cho elaborated these features and defined three criteria useful in identifying the internal opening of a fistula. Criterion I is an appearance of a root-like budding formed by the

D. Cuicchi (✉) · G. Poggioli
Surgery of the Alimentary Tract, Department of Digestive Medicine & Surgery, IRCCS Azienda Ospedaliero Universitaria di Bologna, Alma Mater Studiorum University of Bologna, Bologna, Italy
e-mail: dajana.cuicchi@aosp.bo.it; gilberto.poggioli@unibo.it

M. Lima, G. Ruggeri (eds.), *Ano-Rectal Endosonography and Manometry in Paediatrics*,
https://doi.org/10.1007/978-3-030-97668-2_9

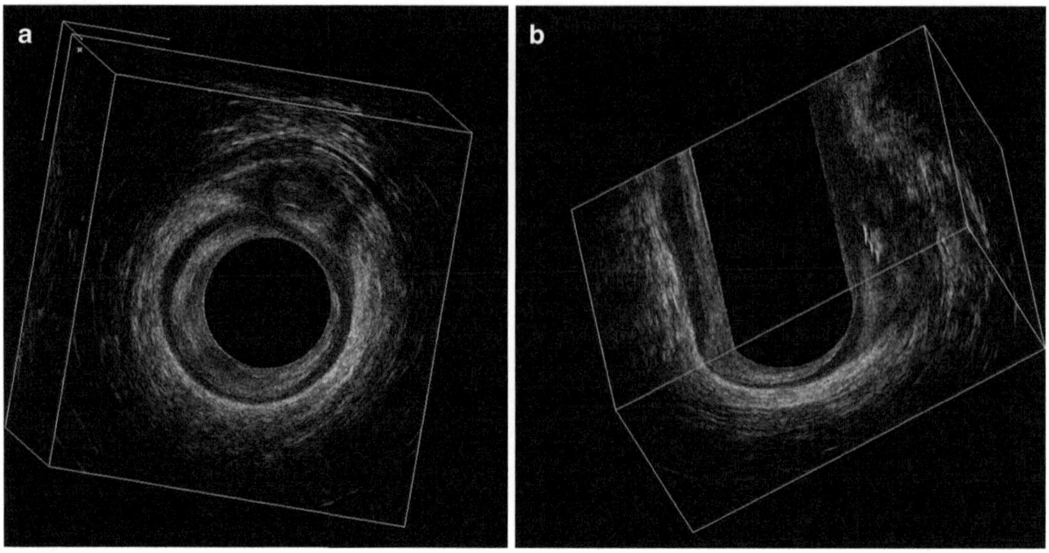

Fig. 9.1 Appearance of perianal fistula at endoanal ultrasound (white arrow): (**a**) transverse endoanal sonographic image in mid anal canal; (**b**) sagittal plane

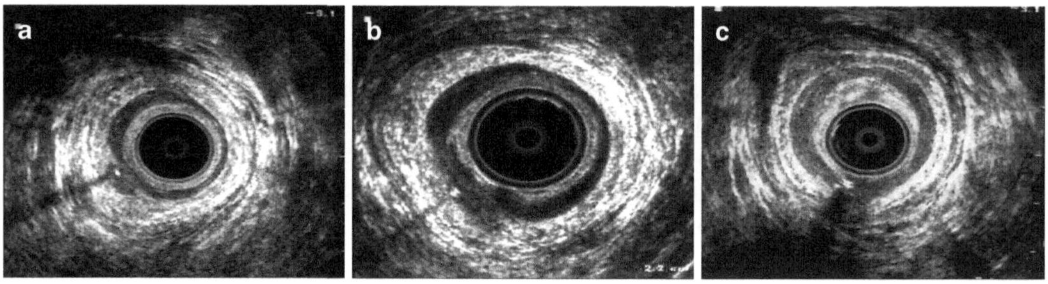

Fig. 9.2 Cho's criteria: (**a**) Criterion I; (**b**) criterion II; (**c**) criterion III

intersphincteric tract which contacts the internal sphincter; criterion II is the appearance of a root-like budding with an internal sphincter defect; criterion III is a subepithelial breech connecting to the intersphincteric tract through an internal sphincter defect (Fig. 9.2) [11]. By using a combination of these three criteria, the sensitivity is 94%, specificity 87%, positive predictive value 81% and negative predictive value 96% [11]. The radial location of the internal opening should be indicated using a clock-face description specifying the patient's position (e.g. Sims or Jackknife position). Longitudinal localisation with respect to the dentate line is not possible because it is not a visible structure at ultrasound; nevertheless, localisation in the low, medium or high anal canal should be reported. Any secondary extensions should always be sought because missed tracks are the main cause of recurrence after treatment. Secondary extensions include abscess collections and secondary blind-ending tracts. Abscess collections appear as hypo-anechoic areas (Fig. 9.3). Their localisation in the perianal, intersphincteric, ischiorectal or suprasphincteric spaces should be defined. The secondary blind-ending tracts are classified as infralevator, supralevator or horizontal extensions, also "horseshoes", according to their relationship to the levator plate (Fig. 9.4). The diagnostic accuracy of EAUS ranges from 63 to 94% for classification of the primary tract and is as high as 93% for the prediction of the site of the internal opening [12–15].

The advent of hydrogen peroxide (H_2O_2) fistulography improved the level of accuracy of EAUS

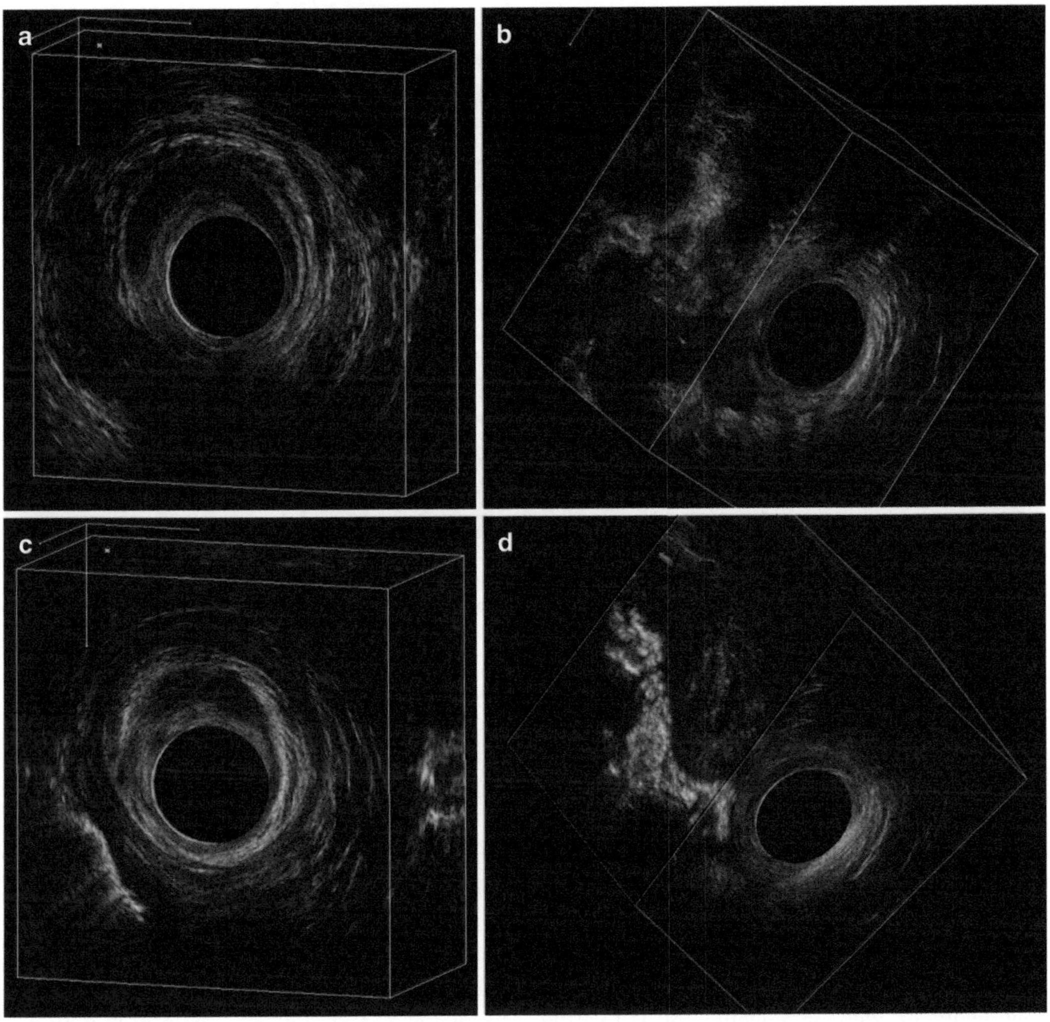

Fig. 9.3 Appearance of abscess collections in right ischiorectal space: (**a**) transverse endoanal sonographic image in mid anal canal at unenhanced endoanal ultrasound; (**b**) sagittal image at unenhanced endoanal ultrasound; (**c**) transverse endoanal sonographic image in mid anal canal at H_2O_2 endoanal ultrasound; (**d**) sagittal image at H_2O_2 endoanal ultrasound

in the characterisation of anal fistulas and allowed easier identification of the tract of complex fistulas. With this technique, described by Cheong et al. in 1993, all the external openings of the fistula were cannulated, and hydrogen peroxide was injected [16]. Peroxide fistulography can assist in the delineation of the internal opening as well as in the detection of the primary tract or extensions as patent fistula tracts will change from hypoechoic to bright hyperechoic due to the presence of multiple hyperechoic gas bubbles within the lumen. A hypoechoic tract which does not show filling with peroxide could represent a fibrotic and inactive tract; however, care must be taken to ensure that all possible external openings have been cannulated and an appropriate length of time has been observed before arriving at this conclusion. Studies of EAUS utilising peroxide fistulography have reported the improvement of results over unenhanced endosonography, with accuracy rates of up to 95% for the classification of the primary tract [17–20]. Poen et al. com-

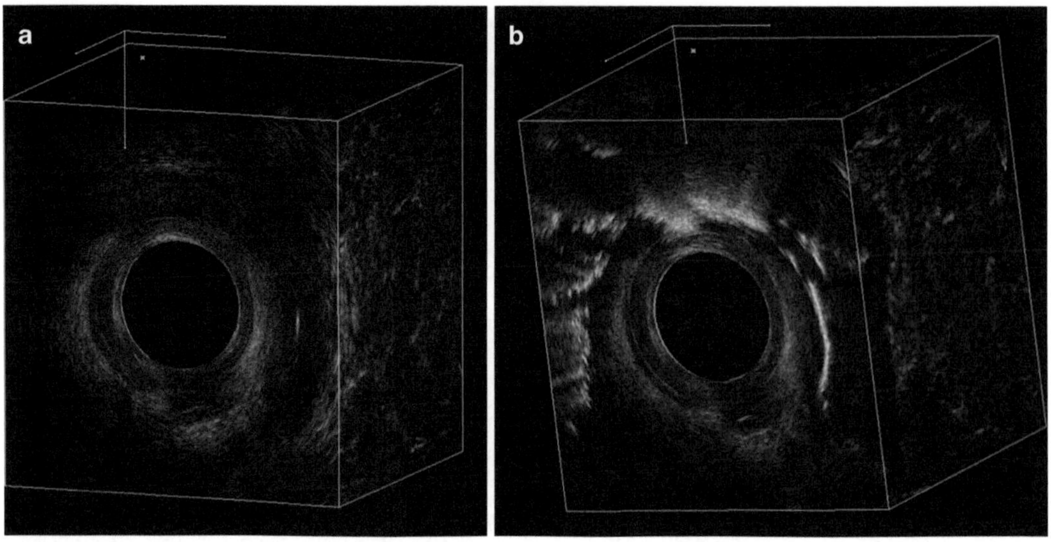

Fig. 9.4 Appearance of horizontal secondary tract, also as "horseshoes": (**a**) transverse endoanal sonographic image at unenhanced endoanal ultrasound; (**b**) transverse endoanal sonographic image at H_2O_2 endoanal ultrasound

pared the intraoperative findings in 21 patients with the results of a clinical examination, a standard US scan and a US scan with H_2O_2 [17]. With the use of contrast medium, the fistula tract was identified in 95% of the patients as compared with 62% at a standard US scan and 38% at a clinical examination. Ratto et al. found that H_2O_2-enhanced EAUS improved the identification of secondary extensions, particularly horseshoe tracks [18]. They performed EAUS, either conventionally or with an injection of H_2O_2; the accuracy rates of EAUS and H_2O_2-enhanced ultrasound were 65.4 and 88.5% for secondary tracts and 80.8 and 92.3% for horseshoe extensions, respectively. In a more recent study, Navarro-Luna et al. compared surgical findings with the results of endoanal ultrasound with contrast medium in 80 patients with complex or recurrent fistulas [20]. The percentages of agreement as regards the identification of the level of the fistula, identification of the internal opening and detection of chronic fistula tracks were 91%, 85% and 75%, respectively. The authors concluded that EUS with H_2O_2 performed by colorectal surgeons with appropriate experience made it possible to achieve excellent results in the preoperative assessment of anal fistulas. Levovist has also been used to enhance the visualisation of fistula tracts and, although the technique is not in widespread use at the present time, it was better at assessing an anal fistula than physical examination and conventional ultrasound [21].

The use of three-dimensional (3D) endoanal sonography with a post-processing modality (volume rendering) has recently been shown to additionally improve overall accuracy in comparison with conventional EAUS [15, 22, 23]. The ability to view reconstructed images in the coronal and sagittal plane is helpful in both displaying the path of the fistula tract and defining the length of the internal and external anal sphincter involvement in order to plan appropriate surgical therapy [24]. Brillantino et al. found a very good correlation between 3D-EAUS and surgical assessment in the classification of the primary fistula tract ($k = 0.93$) and the identification of the fistula internal opening ($k = 0.71$) with an overall sensitivity and specificity of 98.3% and 91.3%, respectively [25]. Several studies have reported greater accuracy of 3D-EAUS over 2D-EAUS in the definition of fistula tracts, with values which reach 98–100% [26–28]. Buchanan et al. reviewed the diagnostic value of hydrogen peroxide in differentiating recurrent and complex fistulas; they did not find any statistically significant differences between

3D EAUS with and without H_2O_2 as regards the identification of the internal opening, level of the fistula tract and secondary extensions. Nevertheless, the authors pointed out that hydrogen peroxide improved the conspicuity of some tracks and internal openings, and might therefore be helpful in difficult cases [23].

Endoanal ultrasound is also helpful in detecting occult sphincter defects which may exist in the case of the recurrence of an anal fistula. This is important to know before planning surgical treatment as these occult defects may turn into symptomatic defects with faecal incontinence post-operatively. Emile et al. assessed the diagnostic utility of 3D-EAUS in the preoperative evaluation of 131 primary and recurrent anal fistulas, and its role in detecting associated anal sphincter defects [29]. The accuracy and sensitivity of EAUS in recurrent anal fistulas were insignificantly lower than those in primary cases (87 vs. 100% for detecting secondary branches); endoanal ultrasound detected occult anal sphincter defects in 5.3% of the patients studied.

In Crohn's disease, the majority of the fistulas are complex. A study involving 41 patients, assessing fistula classification using 3D-H_2O_2 EAUS, showed that only 22% of the fistulas were single intersphincteric or transsphincteric; the remaining 78% were complex: suprasphincteric or extrasphincteric in 12% of cases, secondary extension in 34% of cases and anovaginal in 32% of cases [30]. In Crohn's disease, EAUS is a useful tool for diagnosing perianal fistulas, for differentiating them from cryptoglandular fistulas and for guiding their treatment. To differentiate between Crohn's-related and cryptoglandular fistulas, some ultrasound features have been suggested, such as a specific Crohn ultrasound fistula sign (CUFS), a wider fistula tract (greater than 3 mm), the presence of debris (hyperechoic secretions) in the fistula tract or abscess, and a double or bifurcated fistula tract (Fig. 9.5) [31–33]. The CUFS is defined as a hypoechogenic fistula tract surrounded by a well-defined hyperechogenic area with extension into the perianal tissue, having a thin regular hypoechogenic edge. It can differentiate between Crohn's-related and cryptogenic fistulas with a positive and negative predictive value of 87% and 93%, respectively [31]. In a study of 158 patients, of whom 33 had a diagnosis of Crohn's disease, the maximum width of the fistula tract >4 mm was highly accurate for differentiating between cryptogenic and Crohn's disease fistulas (area under the receiver operating

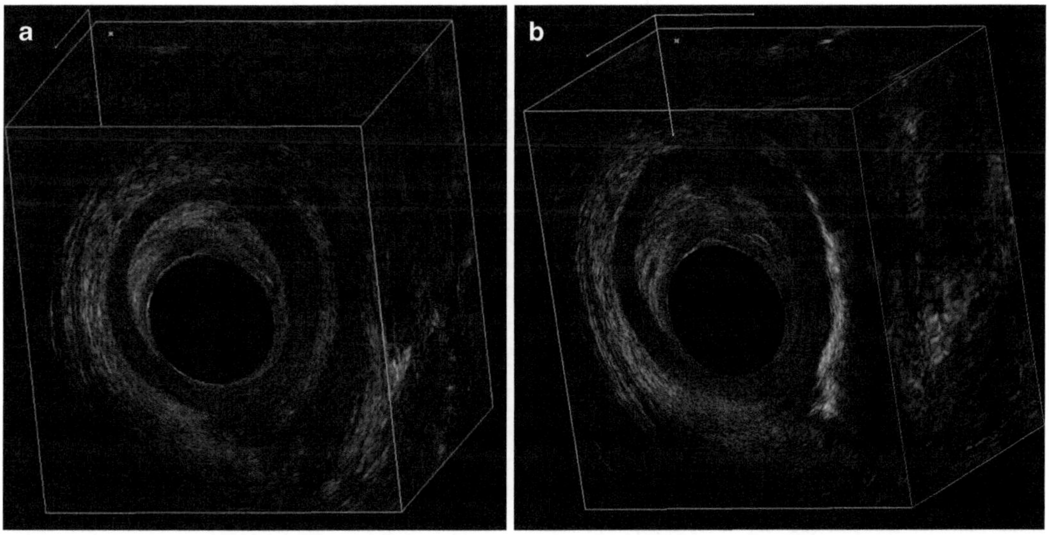

Fig. 9.5 Appearance of horizontal secondary tract of Crohn's-related fistula with Crohn ultrasound fistula sign, wider and bifurcated fistula tract and presence of debris: (**a**) transverse endoanal sonographic image at unenhanced endoanal ultrasound; (**b**) transverse endoanal sonographic image at H_2O_2 endoanal ultrasound

characteristic curve = 0.922). The simultaneous presence of two features raised the probability of having a Crohn's disease fistula to >80%. In particular, the presence of a tract width >4 mm in conjunction with either a double tract or CUFS showed very high specificity (1.00) [34]. In this setting, Crohn's disease should be confirmed on the basis of a full colonoscopy. On the other hand, these features are also useful in identifying cryptoglandular fistulas in patients with established Crohn's disease presenting with a new-onset perianal fistula in order to avoid unnecessary changes in the treatment plan. Endoanal ultrasound may influence patient management in a high percentage of cases [35]. Several randomised prospective studies have shown that EAUS could be successfully used to guide the medical and surgical therapy of perianal fistulising Crohn's disease [36–39]. Endoanal ultrasound is also useful in monitoring the response to treatment with monoclonal drugs and/or antibiotics. During treatment, the healing of the external orifice sometimes precedes fistula tract healing which contributes to the formation of abscesses and fistula recurrence after the discontinuation of treatment [38–42]. Endoanal ultrasound can identify those patients who, presenting healing of the fistula tract, can discontinue treatment without recurrence, and it would therefore be an excellent technique for following these fistulas rather than using magnetic resonance imaging (MRI) each time. In Crohn's disease, pelvic MRI is generally considered to be the initial procedure for assessing a perianal fistula according to the latest available guidelines [43, 44]. On the other hand, 3D-EAUS with H_2O_2 is a good alternative [45].

Several comparative studies have shown the almost identical performance of EAUS and MRI in the detection of anal fistulas [46–48].

A meta-analysis of four studies comparing EAUS and MRI for the assessment of perianal fistulas indicated that the sensitivities of MRI and EUS were comparable (87% and 87%, respectively) and the specificity for MRI was higher than that for EAUS (69% vs. 43%) [49]. The advantages of MRI are its ability to identify high transsphincteric, extrasphincteric and suprasphincteric tracts, and differentiate a recurrent fistula from a post-operative scar [50]. Its use, however, is still limited by cost and access restrictions. Endoanal ultrasound is widely available in the clinical setting; it is an easy-to-use, cost-effective and non-time-consuming imaging modality. However, it has some limitations; it is operator dependent and it has a restricted field of view, reducing the ability to visualise infections which extend far from the focal length of the ultrasound probe, and the use of endoluminal probes can be restricted by luminal stenosis. Both EAUS and MRI provide unique benefits which support a complementary role. The choice of the appropriate initial imaging modality mainly depends on local availability, expertise and complexity of the perianal fistula. To ensure diagnostic accuracy and to determine an optimal management strategy, a combination of diagnostic modalities is recommended. In a prospective study of 32 patients with perianal Crohn's fistulas, the accuracy of each diagnostic modality was excellent (EAUS 91%, MRI 87%); examination under anaesthesia (EUA) was 91% whereas 100% accuracy was achieved with a combination of EUA together with MRI or EAUS [51].

In summary, EAUS is a highly accurate tool for the assessment of both cryptoglandular and Crohn's disease perianal fistulas. It is a simple method, and a rapidly performed, real-time, inexpensive, safe and widely available imaging modality. It is portable and thus potentially available during fistula treatment in the operating theatre. For all these reasons, it should be considered the first examination to be carried out to identify perianal sepsis and in expert hands it also plays an important role in the assessment of complex and recurrent fistulas. It is useful in the preoperative planning of appropriate surgical therapy, decreasing the risk of anal incontinence and recurrence. When deeper abscesses or fistulas are suspected, another MRI should be performed.

References

1. Goligher J. Fistula-in-ano. In: Goligher J, editor. Surgery of the anus, rectum, and colon. 5th ed. Bailliere Tindall; 1984. p. 178–220.

2. Sainio P. Fistula-in-ano in a defined population. Incidence and epidemiological aspects. Ann Chir Gynaecol. 1984;73:219–24.
3. Parks AG, Gordon PH, Hardcastle JD. A classification of fistula-in-ano. Br J Surg. 1976;63:1–12.
4. Rosa G, Lolli P, Piccinelli D, et al. Fistula in ano: anatomoclinical aspects, surgical therapy and results in 844 patients. Tech Coloproctol. 2006;10:215–21.
5. Law PJ, Bartram CI. Anal endosonography: technique and normal anatomy. Gastrointest Radiol. 1989;14:349–53.
6. Law PJ, Talbot RW, Bartram CI, et al. Anal endosonography in the evaluation of perianal sepsis and fistula in ano. Br J Surg. 1989;76:752–5.
7. Nuernberg D, Saftoiu A, Barreiros AP, et al. EFSUMB recommendations for gastrointestinal ultrasound part 3: endorectal, endoanal and perineal ultrasound. Ultrasound Int Open. 2019;5:E34–51.
8. Vogel JD, Johnson EK, Morris AM, et al. Clinical practice guideline for the Management of Anorectal Abscess, fistula-in-Ano, and rectovaginal fistula. Dis Colon Rectum. 2016;59:1117–33.
9. Bartram C, Buchanan G. Imaging anal fistula. Radiol Clin N Am. 2003;41:443–57.
10. Choen S, Burnett S, Bartram CI, et al. Comparison between anal endosonography and digital examination in the evaluation of anal fistulae. Br J Surg. 1991;78:445–7.
11. Cho DY. Endosonographic criteria for an internal opening of fistula-in-ano. Dis Colon Rectum. 1999;42:515–8.
12. Santoro GA, DiFalco G. Endoanal ultrasonography in the evaluation of perianal sepsis and fistula-in-ano: introduction. In: Santoro GA, DiFalco G, editors. Benign anorectal diseases: diagnosis with Endoanal and Endorectal ultrasound and new treatment options. New York: Springer; 2006. p. 130–9.
13. Lengyel AJ, Hurst NG, Williams JG. Pre-operative assessment of anal fistulas using endoanal ultrasound. Color Dis. 2002;4:436–40.
14. Lindsey I, Humphreys MM, George BD, et al. The role of anal ultrasound in the management of anal fistulas. Color Dis. 2002;4:118–22.
15. Santoro GA, Fortling B. The advantages of volume rendering in three-dimensional endosonography of the anorectum. Dis Colon Rectum. 2007;50:359–68.
16. Cheong DM, Nogueras JJ, Wexner SD, et al. Anal endosonography for recurrent anal fistulas: image enhancement with hydrogen peroxide. Dis Colon Rectum. 1993;36:1158–60.
17. Poen AC, Felt-Bersma RJ, Eijsbouts QA, et al. Hydrogen peroxide enhanced transanal ultrasound in the assessment of fistula-in-ano. Dis Colon Rectum. 1998;41:1147–52.
18. Ratto C, Gentile E, Merico M, et al. How can the assessment of fistula-in-ano be improved? Dis Colon Rectum. 2000;43:1375–82.
19. Kruskal JB, Kane RA, Morrin MM. Peroxide-enhanced anal endosonography: technique, image interpretation, and clinical applications. Radiographics. 2001;21:S173–89.
20. Navarro-Luna A, Garcia-Domingo MI, Rius-Macias J, et al. Ultrasound study of anal fistulas with hydrogen peroxide enhancement. Dis Colon Rectum. 2004;47:108–14.
21. Chew SS, Yang JL, Newstead GL, et al. Anal fistula: Levovist-enhanced endoanal ultrasound: a pilot study. Dis Colon Rectum. 2003;46:377–84.
22. Gravante G, Giordano P. The role of three-dimensional endoluminal ultrasound imaging in the evaluation of anorectal diseases: a review. Surg Endosc. 2008;22:1570–8.
23. Buchanan GN, Bartram CI, Williams AB, et al. Value of hydrogen peroxide enhancement of three-dimensional endoanal ultrasound in fistula-in-ano. Dis Colon Rectum. 2005;48:141–7.
24. Garcés-Albir M, García-Botello SA, Esclapez-Valero P, et al. Quantifying the extent of fistulotomy. How much sphincter can we safely divide? A three-dimensional endosonographic study. Int J Colorectal Dis. 2012;27:1109–16.
25. Brillantino A, Iacobellis F, SarnoG D, et al. Role of tri-dimensional endoanal ultrasound (3D-EAUS) in the preoperative assessment of perianal sepsis. Int J Color Dis. 2015;30:535–42.
26. Garcés-Albir M, García-Botello SA, Espi A, et al. Three-dimensional endoanal ultrasound for diagnosis of perianal fistulas: reliable and objective technique. World J Gastrointest Surg. 2016;8:513–20.
27. Ratto C, Grillo E, Parello A, et al. Endoanal ultrasound–guided surgery for anal fistula. Endoscopy. 2005;37:722–8.
28. Kołodziejczak M, Santoro GA, Obcowska A. Three-dimensional endoanal ultrasound is accurate and reproducible in determining type and height of anal fistulas. Color Dis. 2017;19:378–84.
29. Emile SH, Magdy A, Youssef M, et al. Utility of Endoanal ultrasonography in assessment of primary and recurrent anal fistulas and for detection of associated anal sphincter defects. J Gastrointest Surg. 2017;21:1879–87.
30. Sloots CEJ, Felt-Bersma RJF, Poen AC, et al. Assessment and classification of fistula-in-ano in patients with Crohn's disease by hydrogen peroxide enhanced transanal ultrasound. Int J Color Dis. 2001;16:292–7.
31. Zawadzki A, Starck M, Bohe M, et al. A unique 3D endoanal ultrasound feature of perianal Crohn's fistula: the 'Crohn ultrasound fistula sign'. Color Dis. 2012;14:e608–11.
32. Blom J, Nyström PO, Gunnarsson U, et al. Endoanal ultrasonography may distinguish Crohn's anal fistulae from cryptoglandular fistulae in patients with Crohn's disease: a cross-sectional study. Tech Coloproctol. 2011;15:327–30.
33. Zbar AP, Horesh N, Bucholtz V, et al. Are there specific endosonographic features in Crohn's patients with perianal fistulae? J Crohns Colitis. 2013;7:490–6.

34. Luglio G, Giglio MC, Rispo A, et al. Diagnostic accuracy of 3-dimensional endoanal ultrasound in identifying perianal Crohn's fistulas. Dis Colon Rectum. 2018;61:931–7.
35. Lahat A, Assulin Y, Beer-Gabel M, et al. Endoscopic ultrasound for perianal Crohn's disease: disease and fistula characteristics, and impact on therapy. J Crohns Colitis. 2012;6:311–6.
36. Wiese DM, Beaulieu D, Slaughter JC, et al. Use of endoscopic ultrasound to guide adalimumab treatment in perianal Crohn's disease results in faster fistula healing. Inflamm Bowel Dis. 2015;21:1594–9.
37. Spradlin NM, Wise PE, Herline AJ, et al. A randomized prospective trial of endoscopic ultrasound to guide combination medical and surgical treatment for Crohn's perianal fistulas. Am J Gastroenterol. 2008;103:2527–35.
38. Schwartz DA, White CM, Wise PE, Herline AJ. Use of endoscopic ultrasound to guide combination medical and surgical therapy for patients with Crohn's perianal fistulas. Inflamm Bowel Dis. 2005;11:727–32.
39. Ardizzone S, Maconi G, Colombo E, et al. Perianal fistulae following infliximab treatment: clinical and endosonographic outcome. Inflamm Bowel Dis. 2004;10:91–6.
40. West RL, Woude van der CJ, Hansen BE, et al. Clinical and endosonographic effect of ciprofloxacin on the treatment of perianal fistulas. Aliment Pharmacol Ther. 2004;20:1329–36.
41. van Bodegraven AA, Sloots CEJ, Felt-Bersma RJF. Endosonographic evidence of persistence of Crohn's disease-associated fistulas after infliximab treatment, irrespective of clinical response. Dis Colon Rectum. 2002;45(1):39–45.
42. Guidi L, Ratto C, Semeraro S, et al. Combined therapy with infliximab and seton drainage for perianal fistulizing Crohn's disease with anal endosonographic monitoring: a single-Centre experience. Tech Coloproctol. 2008;12:111–7.
43. Bemelman WA, Warusavitarne J, Sampietro GM, et al. ECCO-ESCP consensus on surgery for Crohn's disease. J Crohns Colitis. 2018;12:1–16.
44. Gecse KB, Bemelman W, Kamm MA, et al. A global consensus on the classification, diagnosis and multidisciplinary treatment of perianal fistulising Crohn's disease. Gut. 2014;63:1381–92.
45. Kim Y, Park YJ. Three-dimensional endoanal ultrasonographic assessment of an anal fistula with and without H (2)O(2) enhancement. World J Gastroenterol. 2009;15:4810–5.
46. Alabiso ME, Iasiello F, Pellino G, et al. 3D-EAUS and MRI in the activity of anal fistulas in Crohn's disease. Gastroenterol Res Pract. 2016;2016:1895694.
47. Gustafsson UM, Kahvecioglu B, Aström G, et al. Endoanal ultrasound or magnetic resonance imaging for preoperative assessment of anal fistula: a comparative study. Color Dis. 2001;3:189–97.
48. West RL, Zimmerman DDE, Dwarkasing S, et al. Prospective comparison of hydrogen peroxide-enhanced three-dimensional endoanal ultrasonography and endoanal magnetic resonance imaging of perianal fistulas. Dis Colon Rectum. 2003;46:1407–15.
49. Siddiqui MR, Ashrafian H, Tozer P, et al. A diagnostic accuracy meta-analysis of endoanal ultrasound and MRI for perianal fistula assessment. Dis Colon Rectum. 2012;55:576–85.
50. Sudoł-Szopinska I, Kucharczyk A, Kołodziejczak M, et al. Endosonography and magnetic resonance imaging in the diagnosis of high anal fistulae—a comparison. J Ultrason. 2014;4(14):142–51.
51. Schwartz DA, Wiersema MJ, Dudiak KM, et al. A comparison of endoscopic ultrasound, magnetic resonance imaging, and exam under anesthesia for evaluation of Crohn's perianal fistulas. Gastroenterology. 2001;121:1064–72.

Chronic Constipation in Pediatrics: Pathophysiology and Diagnosis

10

Andrea Pession, Davide Leardini,
Edoardo Muratore, and Riccardo Masetti

10.1 Introduction

Constipation is a common problem in childhood, accounting for 3–5 percent of all visits to pediatricians [1–3]. The incidence in children is 14% and the peak prevalence is during the preschool years [4]. Constipation can have both organic and functional causes. The latter represents the larger group and is defined according to the Rome IV Criteria presented in Table 10.1 [5, 6].

Rome IV Criteria allow symptom-based diagnosis, irrespective of invasive and expensive procedures. However, although in >90% of children the underlying etiology is unrecognized, organic causes should always be excluded. Early diagnosis and treatment of chronic constipation are a priority, since almost one-third of affected children continue experiencing symptoms throughout adolescence despite medical treatment.

10.2 Etiology and Pathophysiology

10.2.1 Functional Constipation

Functional constipation is equally common in both sexes and children with diverse socioeconomic and cultural background. The triggering event is thought to be a non-willing withholding behavior, most likely to avoid pain or for social reasons. Video games seem to represent an important cause of withholding. As a result, the reabsorption of water from the colonic and rectal mucosa makes feces harder and progressively more difficult to evacuate. This leads to a vicious cycle of retention, in which the rectum becomes increasingly distended, resulting in overflow incontinence, loss of rectal sensation, and eventually loss of the normal urge to defecate. Children are particularly prone to develop functional constipation during three time periods:

1. After the introduction of cereals and solid food into the diet (6–12 months of age)
2. During toilet training (2–3 years of age)
3. During the start of school (3–5 years of age)

In older children and adolescents, diets low in fiber and high in dairy, eating disorders, school stressors, attention deficit hyperactivity, and autism spectrum disorder may promote or exacerbate constipation. Overall, psychoemotional

A. Pession (✉) · D. Leardini
E. Muratore · R. Masetti
Pediatric Unit, IRCCS Azienda Ospedaliero-Universitaria di Bologna, Bologna, Italy
e-mail: andrea.pession@unibo.it; riccardo.masetti5@unibo.it

M. Lima, G. Ruggeri (eds.), *Ano-Rectal Endosonography and Manometry in Paediatrics*,
https://doi.org/10.1007/978-3-030-97668-2_10

Table 10.1 Rome IV Criteria for the diagnosis of functional constipation in children (adapted from Benninga MA, Faure C, Hyman PE, et al. Childhood Functional Gastrointestinal Disorders: Neonate/Toddler. Gastroenterology 2016; Hyams JS, Di Lorenzo C, Saps M, et al. Functional Disorders: Children and Adolescents. Gastroenterology 2016)

Infants and toddlers up to 4 years old
At least two of the following present for at least 1 month:
Two or fewer defecations per week
History of excessive stool retention
History of painful or hard bowel movements
History of large-diameter stools
Presence of a large fecal mass in the rectum
In toilet-trained children, the following additional criteria may be used:
At least one episode/week of incontinence after the acquisition of toileting skills
History of large-diameter stools that may obstruct the toilet
Children and adolescents (age ≥4 years)
At least two of the following present at least once per week for at least 1 month: In addition symptoms are insufficient to fulfill the diagnostic criteria of irritable bowel syndrome
Two or fewer defecations in the toilet per week
At least one episode of fecal incontinence per week
History of retentive posturing or excessive volitional stool retention
History of painful or hard bowel movements
Presence of a large fecal mass in the rectum
History of large-diameter stools that may obstruct the toilet

background (e.g., toilet training, starting/changing school, family changes) and inadequate nutrition (diet poor in fiber, rich in fats and sugars, sweet drinks) are considered the major triggers for functional chronic constipation. Alteration in rectum motility, enteric nervous system, and gut microbiota are also being increasingly investigated as key aspects in the pathophysiology [7, 8].

10.2.2 Organic Causes

Organic causes are more common among younger infants because most of them are congenital. The main causes are listed in Table 10.2.

Table 10.2 Main causes of chronic constipation in children

Gastrointestinal causes
Celiac disease
Cow's milk or other dietary protein intolerance
Anorectal anomalies (imperforate anus, anteriorly displaced anus)
Medications (opiates, anticholinergics, antidepressants, chemotherapy, aluminum-containing antacids)
Anal fissure
Inadequate fluid intake (during fever or hot weather)
Immobility
Anorexia nervosa
Starvation
Low dietary fiber
Slow-transit constipation
Intestinal obstruction (in neonates, consider atresia, webs, or volvulus)
Small left colon syndrome
Neurogenic causes
Hirschsprung disease
Cerebral palsy
Myelomeningocele
Spinal cord injury
Closed spinal dysraphism (e.g., tethered cord, sacral agenesis, split spinal cord malformation)
Sacral teratoma
Neurofibromatosis
Muscular weakness (may be generalized, as in down syndrome and Duchenne muscular dystrophy, or due to abnormal abdominal musculature, as in prune belly syndrome or gastroschisis)
Pseudo-obstruction (e.g., visceral neuropathies, myopathies, mitochondrial disorders)
Intestinal neuronal dysplasia
Familial or acquired dysautonomia
Internal anal sphincter achalasia
Endocrine and metabolic causes
Cystic fibrosis (with meconium ileus in neonates or distal intestinal obstruction syndrome in older children)
Hypokalemia
Lead poisoning
Vitamin D intoxication
Hypo- or hypercalcemia
Hypothyroidism
Diabetes mellitus
Pheochromocytoma
Multiple endocrine neoplasia type 2B (MEN2B)
Polyuria (leading to dehydration)

Table 10.2 (continued)

Gastrointestinal causes
Juvenile systemic sclerosis (scleroderma) or mixed connective tissue disease
Acute intermittent porphyria

Table 10.3 Frequency of bowel movement in different age groups

Age	Daily frequency (mean)	Weekly frequency
0–5 months, breastfed	2–6 (2.9)	6–40
0–5 months, formula-fed	1–3 (2.0)	5–28
6–24 months	2–3 (1.8)	4–21
2–4 years	1–2 (1.4)	3–14
5–7 years	0–2 (0.9)	1–7
≥8 years	1–2 (1.0)	1–11

10.3 Clinical Evaluation of Patients with Chronic Constipation

The first goal of the evaluation for a child with chronic constipation is to distinguish between functional and organic causes and this could be usually achieved through focused medical history and physical examination.

10.3.1 Medical History

The history should focus on features that suggest functional constipation and alarm signs. Bowel movement is a key aspect to assess. In infants and children suspicions of chronic constipation should be raised by infrequent evacuation, hard small feces, difficult or painful evacuation of large-diameter stools, and sometimes fecal incontinence and encopresis [9, 10]. Stool frequency and consistency may vary across child ages, so it is important to know expected normal patterns related to age and diet (Table 10.3) [11]. Ninety percent of healthy newborns pass meconium within the first 24 h of life and could be considered normal up to 36–48 h from birth. Premature babies may pass meconium later. During the first 3 months of life, the frequency of bowel movements is influenced by the type of formula used for feeding the infant [12–15]. Breastfed infants may have as few as one stool daily during the first few days of life, and then the frequency increases as mother's milk production increases, up to a mean of three stools per day, but with high interindividual variability. Some breastfed newborns may pass stool with each feeding or may not have a bowel movement any more often than every 7 days. Formula-fed infants pass a mean of two stools per day, but there is variation between formulas. After weaning, the mean number of bowel movements falls progressively to two per day at age two and one per day at age four. The gradual decrease in bowel movement frequency with advancing age correlates with changes in transit time and varying patterns of colonic motility. Other than evaluating constipation with age-adjusted quantitative parameters, qualitative analysis should also be performed using the Bristol Stool Chart. When functional constipation is suspected, particular attention should be given to psychosocial or environmental factors [16]. The history should seek to determine the point at which the constipation was first noted and any potential relationship with an event or factors that may have triggered the constipation. For some individuals, introduction of or increase in cow's milk appears to be a dietary trigger [17]. Other dietary factors that may contribute to constipation include low fiber content (few fruits or vegetables) and low fluid intake. Findings supporting a functional etiology include:

- Onset of constipation coinciding with dietary change, toilet training, painful evacuation, or psychosocial stressor
- Stool-withholding behavior
- Good response to conventional laxatives

History should screen for factors that may suggest an occult organic cause, particularly for younger infants and children presenting with atypical features. Risk for lead poisoning, developmental history (developmental delay may be associated with hypothyroidism and some mitochondrial or neurologic disorders), presence of hypothyroidism, celiac disease, and neuromuscular disorders in the family should be directly investigated. Anorectal and lower urinary tract functions are interrelated, and therefore bladder function must be evaluated in a pediatric patient presenting with chronic constipation. Although bladder dysfunctions, including over- and underactivity and frequent infection, are commonly associated with functional constipation, neurogenic disorders must also be carefully excluded.

"Alarm signs" of a potential organic cause include [1, 18]:

- Delayed passage of meconium (meconium passed after 48 h of life)
- Rectal bleeding (unless attributable to an anal fissure)
- Severe abdominal distension
- Constipation present from birth or early infancy
- "Ribbon" stools (very narrow in diameter)
- Urinary incontinence or bladder disease
- Weight loss, poor weight gain, delayed growth (e.g., decreasing height percentiles)
- Extraintestinal symptoms (especially neurologic deficits)
- Physical findings suggesting possible anorectal anomalies
- Congenital anomalies or syndromes associated with Hirschsprung disease (e.g., genitourinary anomalies, Down syndrome)

10.3.2 Physical Examination

A general physical examination should be performed in all patients. A more detailed focus on the abdomen and perianal area, including the appearance and location of the anus, and sensory and motor function is important in children presenting with chronic constipation. In particular, the physical examination should include assessment of growth, abdominal distension, and abdominal or pelvic masses. It should also evaluate for features suggesting for occult spinal dysraphism, such as increased pigmentation, vascular nevi, or hair tufts in the sacrococcygeal area. The neurologic evaluation should focus on symptoms and signs suggesting spinal cord and/or autonomic nervous system dysfunction, including sensory loss or motor weakness, absent cremasteric reflex, bladder dysfunction, abnormal muscle tone and/or deep tendon reflexes, truncal hypotonia, and altered anal sphincter tone. The perineum should be inspected for abnormalities of anorectal development, which represent a spectrum from high imperforate anus to anteriorly displaced anus [19]. A digital rectal examination is not routinely indicated, but is suggested for the following selected groups of patients [1]:

- Infants with constipation
- Children with symptoms since early infancy
- Infants or children with other alarm signs that suggest organic disease
- Children in whom the presence or degree of constipation is unclear (e.g., meeting only one Rome IV Criterion)

Findings suggestive of Hirschsprung disease include a tight anal canal with an empty ampulla. Explosive release of gas and stool after the digital rectal examination could also occur. In addition, infants with Hirschsprung disease often have gross distension of the abdomen. Findings suggestive of functional constipation are a distended rectum that is full of stool. However, lack of stool does not exclude the possibility of functional constipation. Testing of the stool for occult blood should be performed in most cases if stool is available from the digital rectal examination or diaper. This is particularly important in infants with constipation, in whom subclinical milk protein intolerance (or other food protein intolerance) could present as constipation.

10.4 Laboratory Test, Imaging, and Motility Testing

If warning signs of possible organic constipation are present, focused laboratory and radiographic testing should be performed. In addition, these tests may be appropriate for patients who fail to respond to intervention program.

10.4.1 Blood Tests

For children with failure to thrive, recurrent abdominal pain, or other suspicious signs, a complete blood count (CBC) and serologic screening for celiac disease, usually immunoglobulin A (IgA) antibodies to tissue transglutaminase, are recommended. Urine analysis and urine culture should be requested in children with a history of rectosigmoid impaction and encopresis [20–22]. For children with impaired growth, depressed reflexes, or other signs and symptoms suggesting hypothyroidism, TSH, T4, and T3 should be performed. In patients at risk for electrolyte disturbances (e.g., those with metabolic disease) measuring of serum concentrations of electrolytes and calcium is suggested. Other specific laboratory testing may be considered in any patient with an atypical presentation.

10.4.2 Radiographic Studies

A plain abdominal radiograph is not indicated for the routine evaluation of functional constipation [1]. In selected cases, radiography can be helpful to document retained stool when there is inadequate relevant historical information or if the physical examination is limited by patient cooperation or obesity, or is deferred for psychological considerations. A radiopaque marker study (also known as colon transit study) is generally reserved for the secondary evaluation of selected patients in whom the diagnosis is unclear despite a thorough initial evaluation and trials of treatment [1, 23]. In particular, it may be useful to help distinguish between retentive fecal incontinence (constipation-associated) and non-retentive fecal incontinence [1] and to identify children with "slow-transit" constipation. Children with functional constipation tend to have slower colonic transit time, while a few children with severe slow-transit constipation could have disorders associated with colonic dysmotility (e.g., intestinal neuronal dysplasia). A radiopaque marker study may also help to identify patients with outlet obstruction, manifested by accumulation of markers in the rectosigmoid area. Patients with outlet obstruction may require a biopsy to evaluate for Hirschsprung disease or other neuromuscular disorders [12]. In very young infants, the barium enema may be normal. In children with evidence of spinal dysraphism or neurologic impairment of the perianal area or lower extremities, spine radiographs and/or magnetic resonance imaging should be considered to investigate the possibility of spinal cord tumors and anatomic malformations [1, 24, 25].

10.4.3 Tests for Anorectal and Colonic Function

Anorectal manometry could also be used as the initial investigation or proceed directly to rectal biopsy. Anorectal manometry involves placement of a catheter containing pressure-transducing sensors into the rectum, thereby permitting measurement of neuromuscular function of the anorectum. The procedure includes measurements of the recto-anal inhibitory reflex (which is absent in Hirschsprung disease), rectal sensation and compliance, and squeeze pressures. The test is performed mainly in children with intractable constipation when there is suspicion of internal anal sphincter achalasia or Hirschsprung disease. Anorectal manometry can also identify patients with dyssynergic defecation, which is a functional disorder characterized by the incomplete evacuation of fecal material from the rectum due to paradoxical contraction or failure to relax pelvic floor muscles when straining to defecate. Motility testing should also be considered in patients with no clear organic cause of constipation and/or who fail to respond to regular treatment of functional constipation [26]. However, this pattern may also be seen in patients with fecal impaction and in those with abnormal responses of the pelvic floor muscles during defecation.

10.4.4 Biopsy

Definitive diagnosis of specific pathological conditions can be made by mucosal biopsy, such as Hirschsprung disease of celiac disease [12, 20, 27–29].

References

1. Tabbers MM, DiLorenzo C, Berger MY, et al. European Society for Pediatric Gastroenterology, Hepatology, and Nutrition; North American Society for Pediatric Gastroenterology. Evaluation and treatment of functional constipation in infants and children: evidence-based recommendations from ESPGHAN and NASPGHAN. J Pediatr Gastroenterol Nutr. 2014 Feb;58(2):258–74.
2. van den Berg MM, Benninga MA, Di Lorenzo C. Epidemiology of childhood constipation: a systematic review. Am J Gastroenterol. 2006;101:2401.
3. Robin SG, Keller C, Zwiener R, et al. Prevalence of pediatric functional gastrointestinal disorders utilizing the Rome IV criteria. J Pediatr. 2018;195:134.
4. Mugie SM, Benninga MA, Di Lorenzo C. Epidemiology of constipation in children and adults: a systematic review. Best Pract Res Clin Gastroenterol. 2011;25:3–18.
5. Benninga MA, Faure C, Hyman PE, et al. Childhood functional gastrointestinal disorders: neonate/toddler. Gastroenterology. 2016:S0016-5085(16)00182-7. https://doi.org/10.1053/j.gastro.2016.02.016.
6. Hyams JS, Di Lorenzo C, Saps M, et al. Functional disorders: children and adolescents. Gastroenterology. 2016:S0016-5085(16)00181-5.
7. de Moraes JG, Motta ME, Beltrão et al. Fecal microbiota and diet of children with chronic constipation. Int J Pediatr 2016;2016:6787269.
8. Drossman DA. Functional gastrointestinal disorders: history, pathophysiology, clinical features and Rome IV. Gastroenterology. 2016:S0016-5085(16)00223.
9. Rubin GP. Childhood constipation. Am Fam Physician. 2003;67:1041.
10. Loening-Baucke V. Chronic constipation in children. Gastroenterology. 1993;105:1557.
11. Loening-Baucke V. Prevalence, symptoms and outcome of constipation in infants and toddlers. J Pediatr. 2005;146:359.
12. Di Lorenzo C. Pediatric anorectal disorders. Gastroenterol Clin N Am. 2001;30:269.
13. Fontana M, Bianchi C, Cataldo F, et al. Bowel frequency in healthy children. Acta Paediatr Scand. 1989;78:682.
14. Hyams JS, Treem WR, Etienne NL, et al. Effect of infant formula on stool characteristics of young infants. Pediatrics. 1995;95:50.
15. den Hertog J, van Leengoed E, Kolk F, et al. The defecation pattern of healthy term infants up to the age of 3 months. Arch Dis Child Fetal Neonatal Ed. 2012;97:F465.
16. Arce DA, Ermocilla CA, Costa H. Evaluation of constipation. Am Fam Physician. 2002;65:2283.
17. Iacono G, Cavataio F, Montalto G, et al. Intolerance of cow's milk and chronic constipation in children. N Engl J Med. 1998;339:1100.
18. Leung AK, Chan PY, Cho HY. Constipation in children. Am Fam Physician. 1996;54:611.
19. Hendren WH. Pediatric rectal and perineal problems. Pediatr Clin N Am. 1998;45:1353.
20. Abi-Hanna A, Lake AM. Constipation and encopresis in childhood. Pediatr Rev. 1998;19:23.
21. Hellerstein S, Linebarger JS. Voiding dysfunction in pediatric patients. Clin Pediatr (Phila). 2003;42:43.
22. Loening-Baucke V. Urinary incontinence and urinary tract infection and their resolution with treatment of chronic constipation of childhood. Pediatrics. 1997;100:228.
23. van Summeren JJGT, Holtman GA, van Ommeren SC, et al. Bladder symptoms in children with functional constipation: a systematic review. J Pediatr Gastroenterol Nutr. 2018;67:552.
24. Rosen R, Buonomo C, Andrade R, Nurko S. Incidence of spinal cord lesions in patients with intractable constipation. J Pediatr. 2004;145:409.
25. Siddiqui A, Rosen R, Nurko S. Anorectal manometry may identify children with spinal cord lesions. J Pediatr Gastroenterol Nutr. 2011;53:507.
26. Tambucci R, Quitadamo P, Thapar N, et al. Diagnostic tests in pediatric constipation. J Pediatr Gastroenterol Nutr. 2018;66(4):e89–98.
27. Hussain SZ, Di Lorenzo C. Motility disorders. Diagnosis and treatment for the pediatric patient. Pediatr Clin N Am. 2002;49(27)
28. Osatakul S, Patrapinyokul S, Osatakul N. The diagnostic value of anorectal manometry as a screening test for Hirschsprung's disease. J Med Assoc Thail. 1999;82:1100.
29. Pensabene L, Youssef NN, Griffiths JM, Di Lorenzo C. Colonic manometry in children with defecatory disorders. Role in diagnosis and management. Am J Gastroenterol. 2003;98(1052)

Treatment of Chronic Constipation in Pediatrics

11

Andrea Pession, Davide Leardini, Edoardo Muratore, and Riccardo Masetti

11.1 Introduction

Treatment for chronic constipations differs between functional and organic causes. The latter usually benefit from the treatment of the underlying disease even if sometimes the constipation can persist after the resolution of the pathological conditions. The general approach to treatment of functional constipation depends on the child's age, presence of underlying behavioral or dietary triggers, and chronicity of the symptoms [1]. The goal of therapy is the passage of soft stools, ideally once per day and no less than every other day. The duration of treatment before achieving this goal can last for weeks to months and sometimes years. Parents should be active protagonists in treatment considering the importance of behavioral modification and long-lasting oral therapy. Treatment widely varies between infants and children, also because often chronic constipations in infant can be the result of underlying organic causes (see Chap. 10). Treatment of chronic functional constipation is based on the concept that stool accumulation causes the colon to be unresponsive due to distension. Thus, effective treatment requires complete emptying of the colon before it can become conditioned to work on its own. This is known as "*bowel retraining*" and is based on three mainstays: relief of impaction, maintenance treatment, and patient education including dietary changes [1–3].

A. Pession (✉) · D. Leardini · E. Muratore
R. Masetti
Pediatric Unit, IRCCS Azienda
Ospedaliero-Universitaria di Bologna, Bologna, Italy
e-mail: andrea.pession@unibo.it;
riccardo.masetti5@unibo.it

11.2 Disimpaction

"Fecal impaction" refers to markedly increased amounts of stool in the rectum and colon and is not always present as a finding in patients. Abdominal or digital rectal examination usually can identify the stool mass, even if abdominal radiograph can provide clearer clinical information. Before the initiation of a long-lasting laxative treatment, it is important to "clean out" the feces in order to make effective oral therapy. When an impaction is present specific treatment should be considered. In infants, an enema is usually sufficient to relieve the impaction. Off-the-shelf preparations containing honey and Vaseline are usually available in order to reduce defecation pain [4–6]. For children, disimpaction can be achieved by oral or rectal medication or a combination of both. Oral regimens, usually including polyethylene glycol (PEG), are noninvasive and are particularly valuable for children with a history of painful defecation even if adherence with the necessary volume may be difficult [4, 7]. Rectal medications are more rapidly effective

M. Lima, G. Ruggeri (eds.), *Ano-Rectal Endosonography and Manometry in Paediatrics*,
https://doi.org/10.1007/978-3-030-97668-2_11

and may be a powerful motivator for toilet sitting [8]. Evidence shows that PEG and enemas are equally effective for fecal disimpaction [1].

11.2.1 Oral medications

In most international recommendations PEG with or without electrolytes is indicated as the first-line treatment for initial disimpaction [7, 9]. This is an osmotic laxative and works by causing water to be retained with the stool increasing the number of bowel movements and softening the stool. For outpatients PEG 4000 is generally the most palatable oral laxative and is reasonably well accepted by children [4]. The dose is 1–1.5 g/kg/day by mouth for up to 6 days. The medication should be dissolved in 10 mL/kg body weight of water or flavored beverage. The second-choice medication is usually lactulose which is broken down in the large intestine into mild acids that draw water into the colon, which helps soften the stools. It is usually given once or twice daily. Mineral oil can be another choice for selected children. The recommended dose is 15–30 mL per year of age, up to 240 mL per day. This medication should be avoided in patients with high risk of aspiration pneumonitis [10]. Other oral agents that have been used successfully for disimpaction include magnesium hydroxide, magnesium citrate, sorbitol, senna, and bisacodyl. One withdrawal of oral medications for disimpaction is the higher frequency of fecal incontinency during treatment. In some cases, infants with functional constipation respond to non-pharmacological treatment containing nondigestible osmotically active carbohydrates, such as sorbitol-containing juices (e.g., apple, prune, or pear).

11.2.2 Rectal Medications

For patients with severe impaction, rectal medications may be needed for effective disimpaction [3]. Rectal medications are more rapidly effective than oral medications. For patients with severe impactions, they may help reduce the risk for vomiting. Enema dosages vary with age. Bisacodyl is indicated for older children and the dosage is 5 mL for children between 2 and 10 years and 10 mL for children older than 10 years of age. For infants, glycerin suppositories or saline enemas using a dose of 10–15 mL per kg are generally more recommended [4]. Enemas can be repeated for a more effective disimpaction considering, however, that frequent dosing should be avoided because there are several case reports of life-threatening hyperphosphatemia and hypocalcemic tetany, particularly in young children [11]. Caregiver should be instructed on how to correctly perform enemas. The use of enemas made of soapsuds, tap water, milk and molasses, magnesium, or herbal formulations is not recommended, because of potential complications, which include colitis, water intoxication, bowel perforation, and bowel necrosis.

11.3 Maintenance Therapy

Maintenance therapy should be considered for patients with chronic constipation with no fecal impaction or for those who have successfully completed a disimpaction regimen to “retrain” the bowel and avoid reimpaction. After disimpaction, patients should be promptly treated with a maintenance regimen of oral laxatives. The laxatives that are considered safe and are used most often for children include PEG 3350 (without electrolytes), magnesium hydroxide (milk of magnesia), lactulose, and mineral oil. PEG is the best-studied medication and appears to be safe and more effective compared with lactulose, milk of magnesia, mineral oil, or placebo [1]. The use of PEG with or without electrolytes is thus recommended as the first-line maintenance treatment. A starting dose of 0.4 g/kg/day is recommended and the dose should be adjusted according to the clinical response (0.2–0.8 g/kg/day). Lactulose is considered to be safe for all ages and is recommended in case PEG is not available. The use of milk of magnesia, mineral oil, and stimulant laxatives may be considered as an additional or second-line treatment. Stimulant laxatives, such as senna or bisacodyl, may be helpful for patients with withholding behavior but are generally less indicated because of a more difficulty weaning off. Treatment differs in infants because of the higher prevalence of organic causes and because the safety of pharmacologic interventions is less well established. Lactulose or sorbitol is frequently used in infants at a dose of 1 mL/kg, once or twice daily. The safety of PEG is not well established but in the small reported series treatment at a dose of 0.8 g/kg was effective and no adverse effects were reported [12]. In infants, mineral oil and bisacodyl are not recommended. The parents or caregivers should be instructed in therapy dose adjustment according to the response and to increase the dose every two days until the child has one or two soft stools each day [8]. Maintenance treatment should continue for at least 2 months and all symptoms should be resolved for at least 1 month before discontinuation. Treatment should be decreased gradually and should be stopped once toilet training is achieved. The achievement of this goal can take as long as six months, up to several years. The laxative dose should be gradually decreased to a dose that will prevent fecal incontinence and maintain one to two bowel movements per day. Stopping laxatives too soon will likely lead to a prompt recurrence and disrupt the treatment program. As laxative therapy is discontinued, it is particularly important to reemphasize compliance with the behavioral and dietary regimens because this may help avoid relapses. During the tapering the family should dispose of enemas to rapidly treat relapses.

11.4 Patient Education and Dietary Changes

11.4.1 Behavior Modification

Maintenance therapy should be accompanied by behavior modification. This represents an important part of the therapy because chronic constipations often result in a loss of normal sensation and not a willful act. For children who are already toilet trained modifications consist of a regular toileting regimen with behavioral reinforcement to encourage cooperation. In children who are not fully toilet trained, treatment should be delayed until it has produced effects. Child's family should be educated about the pathogenesis of constipation and the stool withholding to better participate in the behavioral program [4]. Program usually includes the following:

1. *Toilet sitting*: The child should sit on the toilet shortly after the meals for 5–10 min in order to take advantage from the eating simulation. Toilet sitting should occur at the same time each day and should be maintained all over the year, including during vacations. Children should gradually reach a correct sitting position: first, foot should lean on a support in order to raise the knees above the level of the hips. Then, the patient should lean forward and put elbows on his/her knees. Last, bulge of abdomen should be encouraged while straightening the spine. This position is particularly helpful for a child who tends to withhold stool and have an anteriorized anus.
2. *Reward system*: The parents or caregivers should implement a reward system in which the reward is provided for effort rather than success. Retaliatory or punitive measures can result in anger, and resistance to interventions.
3. *Other interventions*: In severe cases, the school and the teacher can be involved in the treatment. Some children may benefit from access to a private bathroom. The teacher should be sensitive to the child's problem, permitting the use of bathroom whenever requested.

11.4.2 Dietary changes

During the maintenance treatment of chronic constipation, it is important to ingest a diet that helps in fecal regularity [4]. Increased intake of fruits and raw vegetables, bran, and whole-grain breads and cereals is commonly recommended, as is adequate intake of fluids other than milk. However, the evidence supporting these interventions is weak, especially in moderate-to-severe constipation.

1. *Fibers*: Children should comply to a balanced diet that includes whole grains, fruits, and vegetables also trying to satisfy personal tastes. For children with acute or mild chronic constipation, a target for dietary fiber can be estimated as the child's age plus 5–10 g/day. Higher doses of fibers can result in constipation. Adequate intake of fiber may be more when laxative therapy is discontinued, to increase the stool bulk and raise the child's awareness of the need to evacuate [3]. Although increasing the intake of fiber is often recommended for acute and chronic constipation, the evidence base for this practice is weak and somewhat conflicting [13]. This may be because dietary fiber can have either beneficial or adverse effects in children with constipation, especially in the setting of chronic constipation and recurrent impactions. Fiber intake, in some children, can result in an excessive colonic distension and in pain, and thus should be highly individualized.
2. *Fluid intake*: Adequate hydration concurs to stool softening and should be encouraged in with chronic constipation. Recommended doses are 1–2 L of water or other nonmilk liquids per day, particularly if they are using fiber supplements. Cow's milk discontinuation can sometimes improve constipation in some children but the level of evidence is low and the reintroduction can be difficult. Milk substitutes, such as soy milk, have discordant results.

11.5 Treatment Failure

During the treatment for chronic constipation patients can experience both treatment failure and relapses. In the former case, the patient training and/or other treatment components should be reviewed, and often an underlying disease is present and further investigations should be conducted. In some cases, behavioral or psychiatric problems can be present, and involvement of a psychologist or behavioral management is recommended [14]. In fact, most treatment failures are caused by inadequate medication or premature discontinuation and thus the treatment should always be analyzed. On the other end, relapses are common even after a proper treatment. For these patients further evaluations and specific treatment should be recommended. Specific evaluation of these patients can show underlying conditions such as internal anal sphincter achalasia or other anatomic causes of constipation, as well as dyssynergic defecation. Appropriate investigations are discussed in Chap. 10. Specific interventions can be indicated for selected patients:

1. *Nasogastric enema*: For inpatients infusion of PEG-electrolyte solution may be considered via nasogastric tube at 25 mL/kg/h up to a maximum of 400 mL/h. Infusion should be modified based on stool emission, vomiting, or abdominal distension.
2. *Stress reduction therapy*: For patients in which stress and anxiety play an important role, relaxation training, stress inoculation, and general stress management procedures can be helpful. Behavioral specialists can be involved in the treatment.
3. *Neurostimulation*: Transcutaneous or sacral implantation represents novel approaches to be considered in children with spinal issues.
4. *Biofeedback therapy*: In specific cases of dyssynergic defecation, biofeedback therapy can be considered to be associated with other therapies. It consists of floor muscle retraining in order to improve bowel function. It is a painless process that uses probes to inform about muscle activity and this information is used to gain sensitivity. However, it requires a highly motivated patient and its efficacy in the pediatric age group is uncertain.
5. *Anal sphincter release*: Some children with internal anal sphincter achalasia may respond to an intervention through injection of botulinum toxin or myectomy. The effect of the botulinum toxin treatment is temporary even if additional injections after few months can be considered [15]. These interventions are effective for patients with the absence of the rectoanal inhibitory reflex on anorectal manometry.

6. *Sacral nerve stimulation*: This intervention can be considered for patients unresponsive to maximal medical treatment even if solid data on pediatric patients are lacking.

References

1. Tabbers MM, DiLorenzo C, Berger MY, et al. Evaluation and treatment of functional constipation in infants and children: evidence-based recommendations from ESPGHAN and NASPGHAN. J Pediatr Gastroenterol Nutr. 2014:258.
2. Abi-Hanna A, Lake AM. Constipation and encopresis in childhood. Pediatr Rev. 1998;19:23.
3. Loening-Baucke V. Encopresis. Curr Opin Pediatr. 2002;14:570.
4. Michail S, Gendy E, Preud'Homme D, Mezoff A. Polyethylene glycol for constipation in children younger than eighteen months old. J Pediatr Gastroenterol Nutr. 2004;39:197.
5. Borowitz SM, Cox DJ, Kovatchev B, et al. Treatment of childhood constipation by primary care physicians: efficacy and predictors of outcome. Pediatrics. 2005;115:873.
6. Tolia V, Lin CH, Elitsur Y. A prospective randomized study with mineral oil and oral lavage solution for treatment of faecal impaction in children. Aliment Pharmacol Ther. 1993;7:523.
7. Bekkali NL, van den Berg MM, Dijkgraaf MG, et al. Rectal fecal impaction treatment in childhood constipation: enemas versus high doses oral PEG. Pediatrics. 2009;124:e1108.
8. Youssef NN, Di Lorenzo C. Childhood constipation: evaluation and treatment. J Clin Gastroenterol. 2001;33:199.
9. Alper A, Pashankar DS. Polyethylene glycol: a game-changer laxative for children. J Pediatr Gastroenterol Nutr. 2013;57:134.
10. Bandla HP, Davis SH, Hopkins NE. Lipoid pneumonia: a silent complication of mineral oil aspiration. Pediatrics. 1999;103:E19.
11. Biebl A, Grillenberger A, Schmitt K. Enema-induced severe hyperphosphatemia in children. Eur J Pediatr. 2009;168:111.
12. Loening-Baucke V, Krishna R, Pashankar DS. Polyethylene glycol 3350 without electrolytes for the treatment of functional constipation in infants and toddlers. J Pediatr Gastroenterol Nutr. 2004;39:536.
13. Chao HC, Lai MW, Kong MS, et al. Cutoff volume of dietary fiber to ameliorate constipation in children. J Pediatr. 2008;45:153.
14. van Dijk M, Benninga MA, Grootenhuis MA, Last BF. Prevalence and associated clinical characteristics of behavior problems in constipated children. Pediatrics. 2010;125:e309.
15. Keshtgar AS, Ward HC, Sanei A, Clayden GS. Botulinum toxin, a new treatment modality for chronic idiopathic constipation in children: long-term follow-up of a double-blind randomized trial. J Pediatr Surg. 2007;42:672.

12 Echo-Assisted Intra-Sphincteric Botulinum Injection

Giovanna Riccipetitoni, Fabrizio Vatta, and Alessandro Raffaele

12.1 Botulinum Toxin: What Is It and How It Works

Botulinum toxin (BT) is a neurotoxin produced by *Clostridium botulinum*. This bacterium produces eight distinguishable exotoxins (A, B, C_1, C_2, D, E, F, and G), with type A being the most potent.

Initially used for the management of strabismus in 1981 [1], BT was then approved for the treatment of several other conditions and it is now used in almost every subspecialty of medicine [2].

BT interferes with neural transmission by blocking the release of acetylcholine at different levels: neuromuscular junction, autonomic ganglia, postganglionic parasympathetic nerve endings, and postganglionic sympathetic nerve endings [3]. By acting at different points, BT may be used for a wide variety of medical conditions, from spastic movement disorders to hyperhidrosis.

BT is expressed in units of biologic activity, where one unit corresponds to the median intraperitoneal lethal dose (LD_{50}) in female Swiss-Webster mice [4].

G. Riccipetitoni (✉) · F. Vatta · A. Raffaele
Department of Pediatric Surgery, Fondazione IRCCS Policlinico San Matteo—University of Pavia, Pavia, Italy
e-mail: g.riccipetitoni@smatteo.pv.it; giovanna.riccipetitoni@unipv.it; a.raffaele@smatteo.pv.it

It takes approximately 4–7 days for BT to reach the peak of the paralytic effect and its effect usually lasts for 8–12 weeks [1]. Around 5–15% of patients do not respond to repetitive injections of BT, due to the production of neutralizing antibodies [5].

Besides being the most potent exotoxin among BT, serotype A is the only one commercially available for clinical use, although experience is emerging in using different serotypes (B, C, and F) [6].

At the moment, among the different preparations, BOTOX® (Allergan, Irvine, California) and Dysport® (Ipsen Ltd., Slough, Berkshire, UK) are the most used and commercialized.

12.2 Perineal Intra-Sphincteric Botulinum Injection

12.2.1 History

The use of BT injections into the anal sphincter was first reported by Langer in 1997. His team started administering BT to decrease obstructive symptoms in patients with Hirschsprung's disease who underwent a pull-through [7]. Since then, BT has been used to treat a variety of other defecation disorders, including functional constipation, internal anal sphincter achalasia, and chronic anal fissures [8].

M. Lima, G. Ruggeri (eds.), *Ano-Rectal Endosonography and Manometry in Paediatrics*,
https://doi.org/10.1007/978-3-030-97668-2_12

12.2.2 How Is It Performed

The procedure is performed under general anesthesia in gynecological position. A rectal touching is performed to appreciate the limits of the internal anal sphincter (IAS). After skin disinfection, a needle (ranging from 21 to 27 gauge) is inserted through the perianal skin and the BT is injected. During the injection, the finger used for the rectal exploration is kept in place, acting as a guide.

The procedure is then repeated circularly around the anus. Different sites of puncture are described: e.g., at 3, 6, and 9 o'clock [9] or 4 injections, one into each quadrant of the internal anal sphincter (Fig. 12.1) [10].

Before each injection the syringe needs to be aspirated to avoid an intravascular injection. Patients are usually discharged from the hospital on the day of the procedure.

Some teams perform the injections transanally using a speculum to identify the dentate line and directly inject into the internal anal sphincter [10–12].

It is important to avoid injecting too superficially, which may cause rectal mucosal ulceration, or too deep into the external sphincter apparatus, resulting in painful inflammation, damage to the sphincter itself, or potential abscess formation [13].

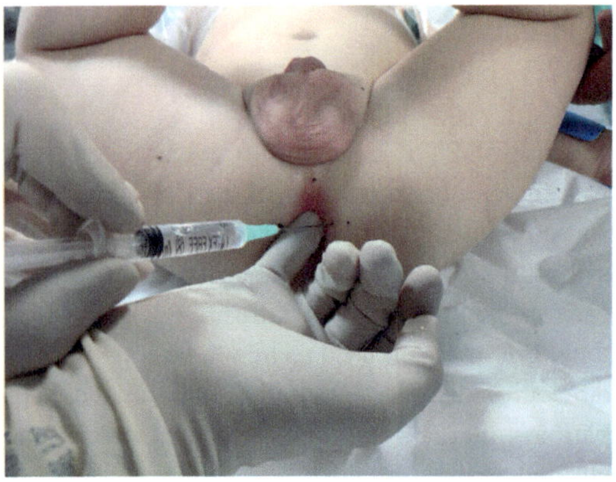

Fig. 12.1 Perineal approach with intra-sphincteric BT injection into four quadrants

12.2.3 Echo-Assisted Injection

An ultrasound probe (honey stick or circular) is inserted into the anus in order to visualize the anal anatomy. The internal anal sphincter can be seen as an anechoic line inferior to the superficial mucosa whereas the external anal sphincter is hyperechoic and is visualized inferiorly to the IAS. BT is then injected under ultrasound visualization into the IAS (Fig. 12.2) [13].

In the last decades many centers started used echo-assisted intra-sphincteric BT injection to treat defecation disorders. In 2017, Church et al. described the use of ultrasound-guided BT injection in patients with Hirschsprung's disease and IAS achalasia. They performed 30 procedures (13 with ultrasound, 17 without), with a median follow-up of 87 weeks. All patients who did not undergo an ultrasound-assisted procedure required a more definitive operation, compared to only 50% in the group of ultrasound-assisted one. The reason for the increased duration of action may be related to a more accurate delivery of the toxin [13]. Ultrasound-assisted BT injection, due to a better visualization, may decrease the absorption in the surrounding tissues, thus providing a higher efficacy [12]. Nevertheless, at the moment, the degree to which this technique decreases complications or increases effectiveness of the BT injection is unknown [13].

12.2.4 Dosage

There is a wide heterogeneity in literature regarding the administered dosage. A large meta-analysis conducted by Roorda et al. in 2019 reports the following dosage: average Dysport® dose of 200 IU per procedure and average Botox® dose of 95 IU per procedure (range 60–200 IU) [14]. Individual Botox® dose of 4–6 IU/kg is reported in literature [9, 10]. BT is diluted to a concentration ranging from 100 IU/mL to 100 IU/5 mL. BT is administered, as already mentioned, at different sites around the anus [9–15]. In our experience, we use 6 IU/kg (maximum dose 100 IU) of Botox®, with a dilution of

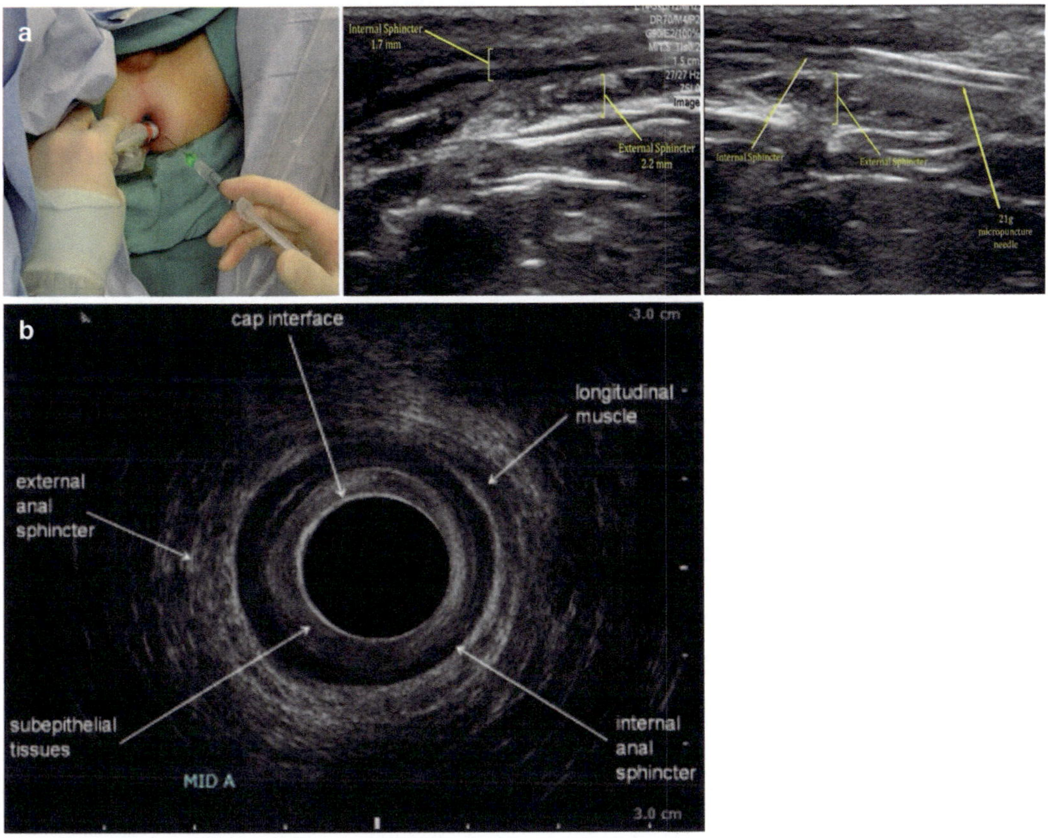

Fig. 12.2 Echo-assisted BT injection. (**a**) BT injection using a hockey-stick probe [courtesy of J.T. Church et al. (2017) [13]]. (**b**) Ultrasound vision using a circular probe [courtesy of R.P. Akbari (2010) [47]]

100 IU in 2 mL. Max 0.5 mL (25 IU) is then injected for each quadrant. It is important to underline that in literature neither dose nor type of BT is predictive of clinical improvement [14, 16].

12.3 Indication

12.3.1 Hirschsprung's Disease

Hirschsprung's disease is a congenital absence of ganglions of the distal gut, causing neonatal bowel obstruction [14]. Although good clinical outcomes are reported after pull-through procedure and resection of the affected aganglionic bowel segment, obstructive gastrointestinal symptoms develop in up to 30% of cases [17, 18]. When treated inadequately, persistent obstructive symptoms may result in fecal stasis that can lead to Hirschsprung-associated enterocolitis, a potential life-threatening condition [19].

Since 1997, when Langer and Birnbaum [7] proposed the use of botulinum toxin to treat these patients, this procedure has gained popularity, as an alternative to myotomy of the anal sphincter or other surgical interventions. In a recent meta-analysis (2019), data from 14 studies, representing 278 patients, were analyzed. BT injections were effective in 66% of patients, ranging from 79% in the first month of follow-up to 46% in follow-up longer than a month. Efficacy was not correlated to age at injection, sex, associated syndromes, dosage, and type of BT used. Enterocolitis incidence was reduced in 57% of cases. On average 2–3 injections per patient were needed to reach a satisfactory clinical improvement [14].

Less satisfactory data were described by Youn et al. in a recent prospective study following 15 patients. They did not report any change in the constipation score, and nonsignificant improvements in the QoL score and resting anal and rectal pressures [9].

12.3.2 Internal Anal Sphincter (IAS) Achalasia

Internal anal sphincter (IAS) achalasia is a condition that shares a similar clinical presentation to Hirschsprung's disease, despite the presence of ganglion cells on rectal suction biopsy (RSB) [20]. The pathophysiology of IAS achalasia is still not fully understood; nevertheless there are several studies demonstrating an altered intramuscular innervation in this condition [21–25]. In patients with IAS achalasia, in fact, there is an absence of nitrergic innervation, which is known to regulate smooth muscle relaxation [22]. Nitric oxide is recognized as a potent mediator of the nonadrenergic noncholinergic (NANC) inhibitory nerves, implicated in the regulation of IAS relaxation in response to rectal distension (recto-sphincteric inhibitory reflex) [20]. The dysfunction of the nitrergic innervation causes thus the failure of relaxation of the IAS, which translates into a clinical presentation of severe constipation with or without soiling [26]. The diagnosis of IAS achalasia is based on anorectal manometry to demonstrate the absence of recto-sphincteric inhibitory reflex, associated with a rectal suction biopsy showing the presence of ganglion cells and normal acetylcholinesterase activity [27]. Although posterior IAS myectomy has long been considered the treatment of choice for this condition [28–30], recent studies show how intra-sphincteric BT injection may be an effective therapeutic alternative in patients with IAS achalasia [31–33]. Friedmacher and Puri performed a meta-analysis with a total of 395 patients, 58% treated with IAS myectomy (M) whereas 42% with BT injection. Improvement after BT was 80.1% (<6 months) and 58.7% (>6 months), compared to 87.8% and 85.2% after M. There was no significant difference in terms of complications (18.8% BT vs. 13.5% M), use of laxatives or rectal enemas (24.1% BT vs. 24.4% M), constipation (20.4% BT vs. 15.5% M), and soiling (3.8% BT vs. 14.3% M). Nevertheless, need for subsequent surgical treatment was more frequent with BT injection (19.2% vs. 4.5%) [34].

12.3.3 Chronic Constipation

Pediatric chronic constipation accounts for about 5% of pediatrician visits and 25% of gastroenterology consultations [35]. Many aspects of its etiology and pathophysiology remain unknown [36], although it seems that a large number of children suffer due to voluntary withholding behavior [37]. The passage of hard stool is associated with pain, leading to fear and anxiety, resulting in a withholding behavior. This mechanism creates a positive feedback that increases stool retention [38]. The proposed treatment involves a combination of medication (e.g., stool softeners), dietary interventions, and behavioral modifications [37] to increase bowel movements with non-painful stool passage. Persistent chronic constipations should be investigated through anorectal manometry and rectal suction biopsy in order to exclude other pathologies, such as IAS achalasia, Hirschsprung's disease, and neuronal dysplasia. In a recent study, C. Zar-Kessler et al. retrospectively analyzed a cohort of 164 children with chronic constipation. Overall response to BT injection was around 70% demonstrating similar response in patients with normal and abnormal sphincters at anorectal manometry. They conclude then that the use of BT should not be limited to patients with atypical sphincter function, but it may also prevent other causes of constipation such as withholding behavior related to pain. Moreover, this study shows how 57% of the cases treated had a beneficial response lasting greater than 6 months, with 17% reporting a positive effect lasting more than 1 year. This suggests how BT injection may serve as a bridge intervention, while waiting as children's symptoms can often improve with time and patient maturity [10].

A preliminary personal experience was presented at the International Pediatric Colo-Rectal

Club in in 2002. Twenty-nine patients were treated and followed for chronic constipation. Median age was 4 years (5 months–16 years). Treatment was effective in 76.7% of children. Among them, the majority improved immediately after the first injection. In 3 cases, improvement was achieved with an associated psychotherapy [39].

12.3.4 Chronic Anal Fissure

An anal fissure is a longitudinal tear in the anal canal. This condition is usually associated with chronic constipation and hypertonia of the IAS. This lesion is defined chronic if it persists for more than 6 or 8 weeks [40]. Lateral internal sphincterotomy is considered to be the gold standard for the treatment, with a healing rate of 88–100% of cases. However, risk of both immediate (8–30%) [41, 42] and long-term [43] incontinence has been reported.

Recent studies have proven that BT is a valid, less invasive alternative to sphincterotomy. In fact, despite a slightly lower healing rate (70–85%), this technique avoids any long-lasting potential complications of IAS sphincterotomy [15, 44].

Another treatment proposed for anal fissure is topical application of pharmacological agents such as nitric oxide donors (e.g., nitroglycerin) and calcium channel blockers (CCB) (e.g., nifedipine, diltiazem) [41]. Topical nitrates have as principal limitation the onset of headache, reported in 20–30% of patients [45]. In a meta-analysis published by Sahebally in 2017, BT injection resulted in better healing rates (71.4 vs. 57.9%) and fewer fissure recurrences (18.5 vs. 25.1%) compared to topical application of nitrates. Nevertheless, a higher rate of transient anal incontinence (10.4% vs. 4.4%) was found in patients treated with BT [46].

12.4 Complications

Intra-sphincteric BT injection is safe and feasible. Side effects, when present, are mild and temporary. Halleran, Levitt et al. [14] performed a literature review analyzing 881 patients (1332 injections), finding an overall complication rate of 0.7% (n = 9), which included temporary urinary incontinence (n = 5), transient pelvic muscle paresis (=2), perianal abscess (n = 1), and rectal prolapse (=1). Other reported complications are bleeding at the site of injection, local tenderness, and swelling. These side effects were temporary and resolved spontaneously [9]. Finally, Roonda et al. reported 9% of transient soiling or incontinence in a cohort of 187 patients [12].

12.5 Personal Series

In our personal experience, we treated 50 patients (27 males and 23 females) with intra-sphincteric BT injection. Median age was 5.2 years (5–15 years). Follow-up ranged from 20 months to 11 years. The procedure was performed under general anesthesia and consisted of four perineal injections, one in each quadrant (individual dose of 6 IU/kg, maximum dose of 100UI, with a dilution of 100 IU in 2 mL, a maximum of 0.5 mL (25 IU) for each quadrant). Macrogol (polyethylene glycol) administration was maintained after treatment for at least 2 months.

Indications for BT injection were chronic constipation (38 cases), IAS achalasia (7 cases), and obstructive symptoms after surgery for Hirschsprung's disease (5 cases). In our series, we achieved an improvement in 78.9% of cases of chronic constipation, in 72% of patients with IAS achalasia, and in 60% of children with Hirschsprung's disease. Need for further injection was higher in the IAS achalasia and in the Hirschsprung groups (respectively in 72% and 100% of cases), compared to the functional constipation group (50%). When needed, treatment was repeated after 3–6 months. Three patients required a subsequent surgical intervention: 1 IAS achalasia and 2 Hirschsprung's disease. The only complication reported was transient soiling in 3% of patients, which regressed spontaneously.

References

1. Scott AB. Botulinum toxin injection of eye muscles to correct strabismus. Trans Am Ophthalmol Soc. 1981;79:734–70.
2. Nigam PK, Nigam A. Botulinum toxin. Indian J Dermatol. 2010;55(1):8–14. https://doi.org/10.4103/0019-5154.60343.
3. Sellin LC. The pharmacological mechanism of botulism. Trends Pharmacol Sci. 1985;6:80–2.
4. Hoffman RO, Helveston EM. Botulinum in the treatment of adult motility disorders. Int Ophthalmol Clin. 1986;26(4):241–50. https://doi.org/10.1097/00004397-198602640-00021.
5. Göschel H, Wohlfarth K, Frevert J, Dengler R, Bigalke H. Botulinum A toxin therapy: neutralizing and nonneutralizing antibodies—therapeutic consequences. Exp Neurol. 1997;147(1):96–102. https://doi.org/10.1006/exnr.1997.6580.
6. Jankovic J, Brin MF. Botulinum toxin: historical perspective and potential new indications. Muscle Nerve Suppl. 1997;6:S129–45.
7. Langer JC, Birnbaum E. Preliminary experience with intrasphincteric botulinum toxin for persistent constipation after pull-through for Hirschsprung's disease. J Pediatr Surg. 1997;32(7):1059–62. https://doi.org/10.1016/s0022-3468(97)90399-7.
8. Basson S, Charlesworth P, Healy C, Phelps S, Cleeve S. Botulinum toxin use in paediatric colorectal surgery. Pediatr Surg Int. 2014;30(8):833–8. https://doi.org/10.1007/s00383-014-3536-4.
9. Youn JK, Han JW, Oh C, Kim SY, Jung SE, Kim HY. Botulinum toxin injection for internal anal sphincter achalasia after pull-through surgery in Hirschsprung disease. Medicine (Baltimore). 2019;98(45):e17855. https://doi.org/10.1097/MD.0000000000017855.
10. Zar-Kessler C, Kuo B, Belkind-Gerson J. Botulinum toxin injection for childhood constipation is safe and can be effective regardless of anal sphincter dynamics. J Pediatr Surg. 2018;53(4):693–7. https://doi.org/10.1016/j.jpedsurg.2017.12.007.
11. Ciamarra P, Nurko S, Barksdale E, Fishman S, Di Lorenzo C. Internal anal sphincter achalasia in children: clinical characteristics and treatment with Clostridium botulinum toxin. J Pediatr Gastroenterol Nutr. 2003;37(3):315–9. https://doi.org/10.1097/00005176-200309000-00020.
12. Halleran DR, Lu PL, Ahmad H, et al. Anal sphincter botulinum toxin injection in children with functional anorectal and colonic disorders: a large institutional study and review of the literature focusing on complications. J Pediatr Surg. 2019;54(11):2305–10. https://doi.org/10.1016/j.jpedsurg.2019.03.020.
13. Church JT, Gadepalli SK, Talishinsky T, Teitelbaum DH, Jarboe MD. Ultrasound-guided intrasphincteric botulinum toxin injection relieves obstructive defecation due to Hirschsprung's disease and internal anal sphincter achalasia. J Pediatr Surg. 2017;52(1):74–8. https://doi.org/10.1016/j.jpedsurg.2016.10.023.
14. Roorda D, Abeln ZA, Oosterlaan J, van Heurn LW, Derikx JP. Botulinum toxin injections after surgery for Hirschsprung disease: systematic review and meta-analysis. World J Gastroenterol. 2019;25(25):3268–80. https://doi.org/10.3748/wjg.v25.i25.3268.
15. Andicoechea Agorría A, Del Casar Lizcano JM, Barbón Remis E, et al. Treatment of a chronic anal fissure with a botulin toxin A injection and fissurectomy. Rev Esp Enferm Dig. 2019;111(9):672–6. https://doi.org/10.17235/reed.2019.6202/2019.
16. Jost S. Chronic anal fissures treated with botulinum toxin injections: a dose-finding study with Dysport(®). Color Dis. 1999;1(1):26–8. https://doi.org/10.1046/j.1463-1318.1999.00006.x.
17. Chumpitazi BP, Nurko S. Defecation disorders in children after surgery for Hirschsprung disease. J Pediatr Gastroenterol Nutr. 2011;53(1):75–9. https://doi.org/10.1097/MPG.0b013e318212eb53.
18. Tariq GM, Brereton RJ, Wright VM. Complications of endorectal pull-through for Hirschsprung's disease. J Pediatr Surg. 1991;26(10):1202–6. https://doi.org/10.1016/0022-3468(91)90335-q.
19. Gosain A, Frykman PK, Cowles RA, et al. Guidelines for the diagnosis and management of Hirschsprung-associated enterocolitis. Pediatr Surg Int. 2017;33(5):517–21. https://doi.org/10.1007/s00383-017-4065-8.
20. Doodnath R, Puri P. Internal anal sphincter achalasia. Semin Pediatr Surg. 2009;18(4):246–8. https://doi.org/10.1053/j.sempedsurg.2009.07.006.
21. Fujimoto T, Puri P, Miyano T. Abnormal peptidergic innervation in internal sphincter achalasia. Pediatr Surg Int. 1992;7:12–7.
22. Hirakawa H, Kobayashi H, O'Briain DS, Puri P. Absence of NADPH-diaphorase activity in internal anal sphincter (IAS) achalasia. J Pediatr Gastroenterol Nutr. 1995;20(1):54–8. https://doi.org/10.1097/00005176-199501000-00010.
23. Kobayashi H, Hirakawa H, Puri P. Abnormal internal anal sphincter innervation in patients with Hirschsprung's disease and allied disorders. J Pediatr Surg. 1996;31(6):794–9. https://doi.org/10.1016/s0022-3468(96)90136-0.
24. Oue T, Puri P. Altered intramuscular innervation and synapse formation in internal sphincter achalasia. Pediatr Surg Int. 1999;15(3–4):192–4. https://doi.org/10.1007/s003830050552.
25. Piotrowska AP, Solari V, Puri P. Distribution of interstitial cells of Cajal in the internal anal sphincter of patients with internal anal sphincter achalasia and Hirschsprung disease. Arch Pathol Lab Med. 2003;127(9):1192–5. https://doi.org/10.1043/1543-2165(2003)127<1192:DOICOC>2.0.CO;2.
26. Lake BD, Puri P, Nixon HH, et al. Hirschsprung's disease: an appraisal of histochemically demonstrated acetylcholinesterase activity in suction rectal biopsy specimens as an aid to diagnosis. Arch Pathol Lab Med. 1978;102(5):244–7.
27. Holschneider AM, Kunst M. Anal sphincter achalasia and ultrashort Hirschsprung's disease. In:

Holschneider AM, Puri P, editors. Hirschsprung's disease and allied disorders. 3rd ed. Berlin: Springer; 2008. p. 297–322.
28. De Caluwé D, Yoneda A, Akl U, Puri P. Internal anal sphincter achalasia: outcome after internal sphincter myectomy. J Pediatr Surg. 2001;36(5):736–8. https://doi.org/10.1053/jpsu.2001.22949.
29. Doodnath R, Puri P. Long-term outcome of internal sphincter myectomy in patients with internal anal sphincter achalasia. Pediatr Surg Int. 2009;25(10):869–71. https://doi.org/10.1007/s00383-009-2436-5.
30. Heikkinen M, Lindahl H, Rintala RJ. Long-term outcome after internal sphincter myectomy for internal sphincter achalasia. Pediatr Surg Int. 2005;21(2):84–7. https://doi.org/10.1007/s00383-004-1332-2.
31. Chumpitazi BP, Fishman SJ, Nurko S. Long-term clinical outcome after botulinum toxin injection in children with nonrelaxing internal anal sphincter. Am J Gastroenterol. 2009;104(4):976–83. https://doi.org/10.1038/ajg.2008.110.
32. Koivusalo AI, Pakarinen MP, Rintala RJ. Botox injection treatment for anal outlet obstruction in patients with internal anal sphincter achalasia and Hirschsprung's disease. Pediatr Surg Int. 2009;25(10):873–6. https://doi.org/10.1007/s00383-009-2438-3.
33. Jiang da P, Xu CQ, Wu B, Li ZZ, Zhang YB, Han FY. Effects of botulinum toxin injection on anal achalasia after pull-through operations for Hirschsprung's disease: a 1-year follow-up study. Int J Color Dis. 2009;24(5):597–8. https://doi.org/10.1007/s00384-008-0591-0.
34. Friedmacher F, Puri P. Comparison of posterior internal anal sphincter myectomy and intrasphincteric botulinum toxin injection for treatment of internal anal sphincter achalasia: a meta-analysis. Pediatr Surg Int. 2012;28(8):765–71. https://doi.org/10.1007/s00383-012-3123-5.
35. Borowitz SM, Cox DJ, Kovatchev B, Ritterband LM, Sheen J, Sutphen J. Treatment of childhood constipation by primary care physicians: efficacy and predictors of outcome. Pediatrics. 2005;115(4):873–7. https://doi.org/10.1542/peds.2004-0537.
36. Mugie SM, Di Lorenzo C, Benninga MA. Constipation in childhood. Nat Rev Gastroenterol Hepatol. 2011;8(9):502–11. Published 2011 Aug 2. https://doi.org/10.1038/nrgastro.2011.130.
37. Tabbers MM, DiLorenzo C, Berger MY, et al. Evaluation and treatment of functional constipation in infants and children: evidence-based recommendations from ESPGHAN and NASPGHAN. J Pediatr Gastroenterol Nutr. 2014;58(2):258–74. https://doi.org/10.1097/MPG.0000000000000266.
38. Nurko S, Zimmerman LA. Evaluation and treatment of constipation in children and adolescents. Am Fam Physician. 2014;90(2):82–90.
39. Riccipetitoni G, Marrello S, LaRiccia A, et al. Botulin toxin in the treatment of internal sphincter hypertonicity: a 3 years follow-up study. Oral presentation at the International Meeting of the Pediatric Colorectal Club, Marseille, France; July 2002.
40. Scholefield JH, Bock JU, Marla B, et al. A dose finding study with 0.1%, 0.2%, and 0.4% glyceryl trinitrate ointment in patients with chronic anal fissures. Gut. 2003;52(2):264–9. https://doi.org/10.1136/gut.52.2.264.
41. Stewart DB Sr, Gaertner W, Glasgow S, Migaly J, Feingold D, Steele SR. Clinical practice guideline for the management of anal fissures. Dis Colon Rectum. 2017;60(1):7–14. https://doi.org/10.1097/DCR.0000000000000735.
42. Simkovic D, Smejkal K, Hladík P. Assessment of sphincterotomy results in patients treated for anal fissure. Rev Esp Enferm Dig. 2000;92(6):399–404.
43. Casillas S, Hull TL, Zutshi M, Trzcinski R, Bast JF, Xu M. Incontinence after a lateral internal sphincterotomy: are we underestimating it? Dis Colon Rectum. 2005;48(6):1193–9. https://doi.org/10.1007/s10350-004-0914-3.
44. Barbeiro S, Atalaia-Martins C, Marcos P, et al. Long-term outcomes of botulinum toxin in the treatment of chronic anal fissure: 5 years of follow-up. United European Gastroenterol J. 2017;5(2):293–7. https://doi.org/10.1177/2050640616656708.
45. Nelson RL. Anal fissure (chronic). BMJ Clin Evid. 2014;2014:0407. Published 2014 Nov 12.
46. Sahebally SM, Meshkat B, Walsh SR, Beddy D. Botulinum toxin injection vs topical nitrates for chronic anal fissure: an updated systematic review and meta-analysis of randomized controlled trials. Color Dis. 2018;20(1):6–15. https://doi.org/10.1111/codi.13969.
47. Akbari RP. Normal endoanal/endorectal ultrasound anatomy and technique. Semin Colon Rectal Surg. 2010;21:191–6. Accessed 10 Dec 2020.

Anorectal Manometry and Endosonography in Paediatric Faecal Incontinence

13

Alireza Safaei Keshtgar

Normal anorectal function and faecal continence are achieved by balance of three interrelated factors including the colonic motility to transport faecal matter, a compliant low-pressure rectum to store the intestinal contents and a high-pressure anal sphincter to keep the content. There are also other contributors to the continence including intact motor and sensory neurological function of anorectum and pelvic floor muscles and psycho-behavioural factors influencing the act of defaecation to intermittently evacuate stool from the rectum.

In a normal child faecal continence develops through a maturation process and toilet training usually between 2 and 4 years of age. Psychological and cultural factors may also influence the timing to achieve voluntary control of defaecation. The primary muscles for faecal continence are the internal anal sphincter (IAS), the external anal sphincter (EAS) and the puborectalis muscle. The smooth muscle of IAS is in a tonic state of contraction that contributes to 50–85% of resting sphincter pressure. The voluntary skeletal muscle of EAS contributes to about 25–30% and the anal cushions account for 5–15% of the resting sphincter pressure [1–5]. In this chapter the role of anorectal manometry and endosonography is discussed to understand the pathophysiology of faecal incontinence in the context of functional overflow incontinence and following surgery for Hirschsprung disease and anorectal malformation.

13.1 Physiology of Defaecation

The process of defaecation begins by distension of the rectum with faecal contents, which leads to transient contraction of the EAS and puborectalis muscles known as 'inflation reflex'. This activity maintains faecal continence whilst simultaneous reflex relaxation of the IAS known as rectoanal inhibitory reflex (RAIR) permits a small sample of faeces to pass into upper sensitive anal canal to discriminate the content as solid, liquid or gas. In adults the first sensation to defaecate occurs when rectal pressure rises to about 18 mm Hg and urge to defaecate happens when the pressure reaches about 55 mmHg [6]. An intact sensory mucosa of anal canal plays an important role in the initiation of defaecation reflex. Further rectal distension elicits stronger and longer duration of reflex relaxation of the IAS and inhibition of the EAS. The rectal content is either contained by voluntary contraction of the EAS or released by relaxation of the muscle depending on the social circumstances to inhibit or to facilitate defaecation. The act of defaecation is also helped by adopting a sitting or squatting position, which

A. S. Keshtgar (✉)
Evelina London Children's Hospital, Guy's and St Thomas' NHS Foundation Trust, London, UK
e-mail: Ali.Keshtgar@gstt.nhs.uk

M. Lima, G. Ruggeri (eds.), *Ano-Rectal Endosonography and Manometry in Paediatrics*,
https://doi.org/10.1007/978-3-030-97668-2_13

straightens the acute anorectal angle, and by Valsalva manoeuvre against a closed glottis and contraction of the abdominal wall muscles, which increases the intra-abdominal and rectal pressure. After defaecation, both the EAS and puborectalis muscles contract and the IAS tone returns to normal by 'deflation reflex', which serves to empty the anal canal and re-establishes the anorectal angle. Thus, gross faecal continence or ability to store large volume of solid or liquid stool is the function of an intact anorectal angle maintained by puborectalis contraction and a compliant low-pressure rectum. However, fine faecal continence or control of small volume of soft stool is the function of coordinated action of the IAS and EAS muscles and an intact anal canal sensation [7]. Therefore, defaecation is a spinal reflex that can be inhibited by voluntary contraction of striated EAS or facilitated by relaxation of the sphincter and contraction of abdominal wall muscles. Normally distension of stomach by food initiates contraction of the colon and rectum and a desire to defaecate known as 'gastrocolic reflex'. As a result of this reflex, children often defaecate about 20–30 min after a meal and the timing of this reflex can be used effectively for toilet training and treatment of childhood constipation.

13.2 Chronic Functional Constipation and Overflow Faecal Incontinence

The term constipation describes a symptom rather than a diagnosis, which is infrequent passage of stool, painful defaecation or difficulty in passing stool. Children have a similar range of frequency of defaecation as adults; however if there is a prolonged interval between passages of hard stool the condition is called obstipation [7, 8]. Chronic childhood constipation is often associated with involuntary passage of fluid or semisolid stool into the clothing called overflow soiling or faecal incontinence. Encopresis is the passage of a normal stool in socially inappropriate places in children over the age of 4, which occurs on a regular basis with no underlying organic cause. This is typically seen in children with emotional problems where they may have secondary gain from their symptoms.

The anorectal manometry and endosonography studies have shown underlying physiological and morphological abnormalities responsible for constipation and faecal incontinence in children. These studies have demonstrated diminished rectal sensation and contraction, abnormal anal sphincter resting pressure, overactivity of the EAS and thickening of the IAS; however there is no consensus if these findings are the primary cause or secondary effect of constipation and faecal retention [9–12]. In normal children, faecal continence is achieved by an effective colorectal peristalsis, a compliant low-pressure rectum to store stool, an adequate anal sphincter pressure and intact sensory and motor neurological functions. The investigations of anorectal manometry, endosonography and colonic transit studies provide an objective diagnosis and assessment of the severity of constipation and faecal incontinence in children. This information enables the clinician to plan a rational treatment strategy and provides valuable information to parents and children to understand the underlying pathophysiology of the disorder.

We have reported 144 children (84 males), median age of 8.1 years (range 3.1–15), with symptoms of chronic constipation and faecal incontinence (Table 13.1) [11]. All children had findings consistent with functional constipation and presence of rectoanal inhibitory reflex (RAIR) on manometry and no evidence of anal sphincter damage on endosonography. The symptoms of overflow faecal incontinence (soiling) were present in 137 (94%) patients, infrequent defaecation of once every 2–3 days or less in 132 (91%) and a palpable faecal mass on abdominal examination in 117 (80%). On anorectal manometry, median size of the rectum was larger than normal 260 mL (60–823) and median amplitudes of rectal and anal sphincter contraction were 3 mmHg (1–25) and 9 mmHg (1–35), respectively. However, the median anal sphincter resting pressure was within the normal range measuring 54 mmHg (19–107) and comparable with reports of healthy children in

Table 13.1 Anorectal manometry findings of normal children and patients with functional constipation

	Age (year)	Anal sphincter resting pressure (mmHg)	Anal canal length (cm)
Normal children			
– Kumar et al. (*n* = 30) [13]	3.9 years (±6 mo)[a]	20 (10–50)[b]	3.0 (2.0–4.0)[b]
– Meunier et al. (*n* = 32) [14]	8.0 years (±3.6 yr)[c]	52.2 (±9.5)[a]	–
Functional constipation and faecal incontinence			
– Keshtgar et al. (*n* = 144) [11]	8.1 years (3.1–15 years)[b]	54 (19–107)[b]	2 (1–4)[b]
– Meunier et al. (*n* = 144) [14]	6.0 years (±3.7 yr)[a]	72.8 ± 28.7	–
– El-Shabrawi et al. (*n* = 50) [17]	7.3 years (6–14 yr)[c]	49.5 (15.4–76.5)[b]	2.8 (2–3.9)[c]

Abbreviations: yr (years), mo (months), d (days)
[a] Denotes mean (standard deviation)
[b] Denotes median (range)
[c] Denotes mean (range)

the literature [13, 14]. On endosonography the median thickness of IAS was 0.9 mm (0.3–2.8) and there was no correlation between IAS thickness and both anal sphincter resting pressure and length of anal canal. However, severity of symptoms and soiling and size of megarectum were significantly correlated to the thickness of the IAS muscle. There was overactivity and high amplitude of the IAS on manometry; however this did not cause an increased anal sphincter resting pressure and obstructive defaecation.

In keeping with our findings other investigators have reported that the RAIR threshold was increased in 6.2% of the constipated patients and there was a decreased rectal sensitivity in 68% of the constipated children [14]. In contrast maximal anal resting pressure was raised in 46% of the constipated children. In another study follow-up manometry findings of patients treated for constipation and soiling showed that mean value of RAIR threshold was comparable with healthy children. In these children the mean measurements of anal resting tone, anal pull-through pressure and percent relaxation of RAIR remained significantly lower in both recovered and non-recovered constipated and encopretic patients compared to controls [15].

In children with chronic functional constipation and megarectum, distension of the rectum by retained faecal mass causes reflex inhibition and relaxation of the IAS followed by involuntary passage of stool and overflow incontinence [16]. On manometry, the IAS relaxes more for longer period and it recovers slower in patients with constipation and soiling compared to constipation alone. We concur with other investigators that first sensation, urge to defaecate, intense urge and maximum tolerable volume are higher in constipated children with faecal incontinence [10, 17]. This is likely related to larger size of the rectum (megarectum) and faecal impaction and higher threshold required to stimulate afferent sensory nerve pathways with consequent reduced sensation of urge to defaecate. However follow-up manometry studies have shown improvement of abnormal manometry findings with increased resting anal sphincter and squeeze pressures after 6 months of treatment, when compared to pre-treatment values [17]. Furthermore, reduced rectal sensations related to megarectum significantly improved and reached near-normal values following treatment. These findings show that abnormalities of anorectal motor and sensory functions are likely to be secondary in nature and reversible following successful treatments of chronic constipation and overflow faecal incontinence.

13.3 Defaecation Disorders and Faecal Incontinence in Hirschsprung Disease

Many patients with Hirschsprung disease have a good outcome following pull-through surgery. Long-term follow-up studies have shown problems in some of the children with constipation and faecal incontinence. There is a wide variation in reported outcome post-operatively because different methods and criteria have been used for the evaluation of bowel function. Frequent bowel actions and passing loose stool are common during early post-operative period, which may lead to perianal skin excoriation, faecal urgency, painful defaecation and subsequent withholding behaviour. Stool frequency improves in more than 80% of children to three or fewer bowel actions per day by 3 years of age [18]. Constipation has been reported in 20–35% of children and faecal incontinence in 6–58% of patients with Hirschsprung disease using different pull-through techniques [19–21]. In some children, the underlying causes of obstructive bowel symptoms may be anal stenosis, large Duhamel pouch, residual or acquired aganglionic bowel segment at anastomosis, intestinal neuronal dysplasia (IND), achalasia of the IAS and withholding behaviour [22–25]. However, faecal incontinence may be secondary to sphincter damage during pull-through or after sphincterotomy for obstructive symptoms, non-compliant neorectum and loss of transitional epithelium due to low anastomosis below dentate line. It is generally acknowledged that bowel function improves with age and most children gain normal bowel function in the long term although this is disputed by some investigators [19, 21]. These problems can be identified by detailed history and clinical examination and biopsy of neorectum. However functional studies consisting of anorectal manometry and intestinal transit studies and imaging of anal sphincter complex by endosonography and MRI provide objective assessment and additional information about the underlying pathophysiology of the problems. This is pertinent following reconstructive surgery for Hirschsprung disease as the IAS is non-relaxing, the rectum is removed and the more proximal large bowel may have dysmotility despite a technically successful operation [26].

In our experience and the setting of paediatric colorectal clinics we have observed that most children with problems following surgery for Hirschsprung disease can be classified as suffering from faecal incontinence, constipation or a combination of both. We have investigated the mechanisms responsible for faecal incontinence and constipation by anorectal manometry, endosonography and colonic transit study following the pull-through surgery. Of 19 children (15 boys and 4 girls), 16 had undergone Duhamel, 1 Rehbein and 2 Soave operation (Table 13.2) [27]. On endoanal sonography, all children had an intact IAS and EAS, below the level of pull-through surgery. The anorectal manometry showed a significantly lower resting anal sphincter pressure in the faecally incontinent children compared to the constipated children with incontinence and those without incontinence with respective measurements of median 38 mmHg versus 57 mmHg or median 38 mmHg versus 66 mmHg (Table 13.3). The rectal pressure was significantly higher in incontinent children compared to those with constipation and soiling or constipation only with respective median values of 71 mmHg versus 42 mmHg or median 71 mmHg versus 36 mmHg, $P < 0.05$. Therefore, the ratio of rectal to anal sphincter pressure was higher in incontinent children due to loss of storage capacity as compared to constipated patients. These findings were in agreement with other investigators that in patients with faecal incontinence mean anal resting pressures and maximum tolerable rectal volume were significantly lower than those children, who were continent with respective measurements of mean 47 mmHg versus 63 mmHg, $P < 0.05$, and mean 97 mL versus 181 mL, $P < 0.05$ (Table 13.2) [28]. Therefore, constipation can be caused by high anal resting pressure and weak rectal peristalsis due to megarectum, whilst faecal incontinence may be secondary to poor rectal compliance and elevated pressure in the presence of normal or low anal sphincter pressure. Faecal incontinence may also be caused by overflow soiling, sphincter damage due to pull-through surgery or anastomotic leakage, IAS myectomy and low anastomosis below dentate line. These patients benefit from bowel management programme including low-residue diet to firm consistency of stool combined with stimulant laxatives and if these fail rectal washout or antegrade colonic enema (ACE) stoma [27, 29]. Children with constipation and overflow

Table 13.2 Anorectal manometry and endosonography findings of patients with Hirschsprung disease and anorectal malformations

	Age (year) (range)	Anal sphincter resting pressure (mmHg)	Internal anal sphincter (IAS) on endosonography
Hirschsprung disease			
– Keshtgar et al. (n = 19) [27]	6 years (3–18.5 yr)[a]		
Faecal incontinence (n = 6)		38 (18–53)[a]	Normal
Faecal incontinence and constipation (n = 8)		57 (23–77)[a]	Normal
Constipation (n = 5)		66 (20–70)[a]	Normal
– Tran et al. (n = 19) [28]	11.3 (±6.3 yr)[b]		
Faecal incontinence (n = 12)		47 (±12)[b]	–
Continence (n = 7)		63 (±11)[b]	–
Anorectal malformations			
– Keshtgar et al. (n = 54) [33]			
Low lesions (n = 20)	9.4 years (4.1–15.3 yr)[a]	44 (31–72)[a]	Localised scar
High lesions (n = 34)	10.5 years (3.9–21.8 yr)[a]	24 (9–45)[a]	Fragmented
– Caldaro et al. (n = 17) [34]			
Low lesions (n = 6)	8.3 years (5–15 yr)[c]	42 (±11)[b]	Localised scar (5)
Intermediate lesions (n = 4)	7.2 years (5–10 yr)[c]	20 (±6)[b]	Defect (2), scar (2)
High lesions (n = 7)	8.1 years (6–12 yr)[c]	14 (±9)[b]	Absent (6)

Abbreviations: yr (years), mo (months), d (days)
[a]Denotes median (range)
[b]Denotes mean (standard deviation)
[c]Denotes mean (range)

soiling respond to laxative stool softener and bowel stimulants. Some of these patients may have fear of defaecation and withholding behaviour because of painful anorectal sensation leading to dyssynergia of defaecation dynamics and abnormal contraction of the EAS muscles [30]. In these children the obstructive symptoms are worse because of achalasia and non-relaxing nature of the IAS muscles leading to megacolon, abdominal distension and enterocolitis. They respond well to intrasphincteric injection of botulinum toxin, which is more effective if the toxin is injected into the EAS muscles [25].

It is imperative to investigate the anorectal function and morphology in children with recalcitrant constipation, faecal incontinence and obstructive defaecation following surgery for Hirschsprung disease at an early stage. This will obviate negative social impact of the problems before the child approaches school or infant school age. The anorectal manometry, endosonography and colonic transit studies should be considered in symptomatic children, who do not respond to conservative treatments and require further assessments including rectal biopsy to exclude residual aganglionic segment and intestinal neuronal dysplasia (IND). These studies enable clinicians to understand the underlying cause and plan a rational treatment before embarking on a major surgery [27].

13.4 Constipation and Faecal Incontinence in Anorectal Malformations

Long-term outcome following surgery for anorectal malformation shows that bowel and bladder dysfunction is prevalent in these patients. There are several factors that contribute to the functional sequelae post-operatively including severity of the anomaly, associated neuropathy and megarectum and surgical technique. In a large series of children who had repair of anorectal malformations an overall assessment showed that 43% had voluntary bowel movement and no

Table 13.3 Anorectal manometry scoring system [33]

	Score[a]
Mean resting pressure >30 mmHg	0
Mean resting pressure <30 mmHg 2	2
Minimum sphincter pressure <50% of resting pressure[b]	0
Minimum sphincter pressure >50% of resting pressure[b]	1
Rectoanal inhibitory reflex present	0
Rectoanal inhibitory reflex absent	1

[a] Scores of 0–1 indicate good and scores of 2–4 indicate poor sphincter quality

[b] Reduction of anal sphincter resting pressure on inflation of rectal balloon

soiling, 48% were constipated and 56% had problems with faecal incontinence. The incidences of faecal incontinence after repair of bladder neck, vaginal and prostatic fistulae were 88%, 80% and 77%, respectively. However, constipation was commoner in less severe type of malformations including perineal, vestibular and bulbar fistulae with respective rates of 57%, 63% and 58% [31].

Factors that may adversely affect the outcome of bowel function and faecal continence include high malformations, severe sacral abnormalities with loss of more than two vertebrae, intraspinal anomalies and absence of a functional IAS [31, 32].

We evaluated the role of anorectal manometry and anal endosonography by devising a scoring system to assess the function and quality of the IAS and faecal continence following repair of anorectal malformations [33]. On manometry, a resting sphincter pressure of 30 mmHg or more and presence of rectoanal inhibitory reflex (RAIR) indicated good function (score 0–1) and lower sphincter pressure and no reflex was poor function (score 2–4) (Table 13.3). On endosonography, an intact or scarred IAS was classified as good quality (score 0–1) and a fragmented or absent IAS as poor quality (score 2–3) (Table 13.2, Chap. 5). Bowel function was assessed by modified Wingfield score with normal function (score 0), constipation (score 1), intermittent soiling more than three episodes per week (score 2) and continual soiling every day (score 3). There were 54 children, 34 with high and intermediate malformations and 20 with low lesions with a perineal opening (Table 13.2) [33]. A variety of techniques were used to perform reconstruction of the malformations by several surgeons from different centres. Of 35 children, who had associated megarectum and neuropathy, 16 required subsequent excision of megarectum. The median anal resting sphincter pressure in children with high malformations was 24 mmHg (range 9–45), which was lower compared to those with low lesions, who had a resting sphincter pressure of 44 mmHg (range 31–72) and a better faecal continence comparable to healthy children. RAIR was seen in 80% of children with low and in 18% of those with high malformations. However, presence of megarectum and neuropathy was associated with poor faecal continence irrespective of the IAS quality. These findings are in agreement with results of other investigators, who have reported a correlation between the presence of a functional IAS and high anal resting pressure and good faecal continence following surgery for anorectal malformations. On the other hand, there is an association between poor faecal continence and absence of a functional IAS, severe sacral anomalies and constipation [32, 34, 35].

Children with neuropathic features like sacral dysgenesis are likely to have impairment of functions of the EAS, pelvic floor and motility and sensation of the rectum. In the context of anorectal malformation, neuropathy affecting the bladder is one of the best indicators of neuropathic bowel because of similar embryological origin of hindgut and bladder [33]. Presence of neuropathy and megarectum is particularly unfavourable because of poor compliance and activity of the rectum and high ratio of rectal to anal sphincter pressure. In addition, dysmotility of the megarectum leads to incomplete evacuation and retention of stool followed by faecal impaction and overflow soiling [36]. Therefore, faecal incontinence may be caused by an impacted megarectum and reflex inhibition of the IAS and consequent low sphincter resting pressure, or it may be caused by deficiency of the IAS. In our experience excision of megarectum may be required in these children to gain faecal continence in the presence of a good IAS and absence of neuropathy. However, those patients who have either poor IAS or neuropathy often required artificial means of faecal continence like formation of an antegrade colonic enema (ACE) stoma.

Evidence from the literature and our results show the importance of preserving the IAS at the time of repair of anorectal malformations [32, 33, 35]. These also highlight the prognostic value of anorectal manometry and endosonography with regard to the evaluation of outcome and objective assessment of anorectal physiology and morphology to achieve faecal continence by implementing the most appropriate treatment strategy [33].

References

1. Duthie HL, Watts JM. Contribution of the external anal sphincter to the pressure zone in the Anal Canal. Gut. 1965;6:64–8.
2. Frenckner B, Ihre T. Influence of autonomic nerves on the internal and sphincter in man. Gut. 1976;17(4):306–12.
3. Gibbons CP, Trowbridge EA, Bannister JJ, Read NW. Role of anal cushions in maintaining continence. Lancet. 1986;1(8486):886–8.
4. Lestar B, Penninckx F, Kerremans R. The composition of anal basal pressure. An in vivo and in vitro study in man. Int J Color Dis. 1989;4(2):118–22.
5. Schweiger M. Method for determining individual contributions of voluntary and involuntary anal sphincters to resting tone. Dis Colon Rectum. 1979;22(6):415–6.
6. Ganong WF. Review of medical physiology. 13th ed. San Francisco: Prentice-Hall International Editions; 1987. 665 p.
7. O'Neill JA Jr, Grosfeld JL, Fonkalsrud EW, Coran AG. Other disorders of the anus and rectum, anorectal function. In: Craven L, editor. Pediatric surgery. 2. 5th ed. St Louis, Missouri: Mosby; 1998. p. 1449–60.
8. Talley NJ, Weaver AL, Zinsmeister AR, Melton LJ 3rd. Functional constipation and outlet delay: a population-based study. Gastroenterology. 1993;105(3):781–90.
9. Hosie GP, Spitz L. Idiopathic constipation in childhood is associated with thickening of the internal anal sphincter. J Pediatr Surg. 1997;32(7):1041–3; discussion 3–4.
10. Keshtgar AS, Choudhry MS, Kufeji D, Ward HC, Clayden GS. Anorectal manometry with and without ketamine for evaluation of defecation disorders in children. J Pediatr Surg. 2015;50(3):438–43.
11. Keshtgar AS, Ward HC, Clayden GS, Sanei A. Thickening of the internal anal sphincter in idiopathic constipation in children. Pediatr Surg Int. 2004;20(11–12):817–23.
12. Loening-Baucke VA. Abnormal rectoanal function in children recovered from chronic constipation and encopresis. Gastroenterology. 1984;87(6):1299–304.
13. Kumar S, Ramadan S, Gupta V, Helmy S, Atta I, Alkholy A. Manometric tests of anorectal function in 90 healthy children: a clinical study from Kuwait. J Pediatr Surg. 2009;44(9):1786–90.
14. Meunier P, Marechal JM, de Beaujeu MJ. Rectoanal pressures and rectal sensitivity studies in chronic childhood constipation. Gastroenterology. 1979;77(2):330–6.
15. Loening-Baucke VA, Cruikshank BM. Abnormal defecation dynamics in chronically constipated children with encopresis. J Pediatr. 1986;108(4):562–6.
16. Keshtgar AS, Ward HC, Clayden GS. Pathophysiology of chronic childhood constipation: functional and morphological evaluation by anorectal manometry and endosonography and colonic transit study. J Pediatr Surg. 2013;48(4):806–12.
17. El-Shabrawi MHHM, Abdelgawad MMAH, Hassanin F, Mahfouze AAA, Khalil AFMESE. High-resolution anorectal manometry in children with functional constipation: a single-Centre experience before and after treatment. Gastroenterology Rev. 2018;13(4): 305–12.
18. Rintala RJP, M. Hirschsprung's disease. In: Stringer MD, Oldham KT, Mouriquand PDE, editors. Pediatric surgery and urology: long term outcome. 2nd ed. New York: Cambridge University Press; 2006. p. 385–400.
19. Catto-Smith AG, Nolan TM, Coffey CM. Clinical significance of anismus in encopresis. J Gastroenterol Hepatol. 1998;13(9):955–60.
20. Moore SW, Albertyn R, Cywes S. Clinical outcome and long-term quality of life after surgical correction of Hirschsprung's disease. J Pediatr Surg. 1996;31(11):1496–502.
21. Yanchar NL, Soucy P. Long-term outcome after Hirschsprung's disease: patients' perspectives. J Pediatr Surg. 1999;34(7):1152–60.
22. Baillie CT, Kenny SE, Rintala RJ, Booth JM, Lloyd DA. Long-term outcome and colonic motility after the Duhamel procedure for Hirschsprung's disease. J Pediatr Surg. 1999;34(2):325–9.
23. Morikawa Y, Matsufugi H, Hirobe S, Yokoyama J, Katsumata K. Motility of the anorectum after the Soave-Denda operation. Prog Pediatr Surg. 1989;24:67–76.
24. Soave F. Endorectal pull-through: 20 years experience. Address of the guest speaker, APSA, 1984. J Pediatr Surg. 1985;20(6):568–79.
25. Thakkar HS, Bassett C, Hsu A, Manuele R, Kufeji D, Richards CA, et al. Functional outcomes in Hirschsprung disease: a single institution's 12-year experience. J Pediatr Surg. 2017;52(2):277–80.
26. Tariq GM, Brereton RJ, Wright VM. Complications of endorectal pull-through for Hirschsprung's disease. J Pediatr Surg. 1991;26(10):1202–6.

27. Keshtgar AS, Ward HC, Clayden GS, de Sousa NM. Investigations for incontinence and constipation after surgery for Hirschsprung's disease in children. Pediatr Surg Int. 2003;19(1–2):4–8.
28. Tran VQ, Mahler T, Bontems P, Truong DQ, Robert A, Goyens P, et al. Interest of anorectal manometry during long-term follow-up of patients operated on for Hirschsprung's disease. J Neurogastroenterol Motil. 2018;24(1):70–8.
29. Langer J. Hirschsprung disease. In: Holcomb III GW, Murphy JP, St. Peter SD, editors. Holcomb and Ashcraft's pediatric surgery. 7th ed. Elsevier Inc.; 2020. p. 557–76.
30. Meinds RJ, Eggink MC, Heineman E, Broens PM. Dyssynergic defecation may play an important role in postoperative Hirschsprung's disease patients with severe persistent constipation: analysis of a case series. J Pediatr Surg. 2014;49(10):1488–92.
31. Pena ALMA. Anorectal malformations: experience with the posterior sagittal approach. In: Stringer MDOKT, Mouriquand PDE, editors. Pediatric surgery and urology: long term outcome. 2nd ed. New York: Cambridge University Press; 2006. p. 401–15.
32. Rintala RJ, Lindahl H. Is normal bowel function possible after repair of intermediate and high anorectal malformations? J Pediatr Surg. 1995;30(3):491–4.
33. Keshtgar AS, Athanasakos E, Clayden GS, Ward HC. Evaluation of outcome of anorectal anomaly in childhood: the role of anorectal manometry and endosonography. Pediatr Surg Int. 2008;24(8):885–92.
34. Caldaro T, Romeo E, De Angelis P, Gambitta RA, Rea F, Torroni F, et al. Three-dimensional endoanal ultrasound and anorectal manometry in children with anorectal malformations: new discoveries. J Pediatr Surg. 2012;47(5):956–63.
35. Rintala R, Lindahl H, Marttinen E, Sariola H. Constipation is a major functional complication after internal sphincter-saving posterior sagittal anorectoplasty for high and intermediate anorectal malformations. J Pediatr Surg. 1993;28(8):1054–8.
36. Keshtgar AS, Ward HC, Richards C, Clayden GS. Outcome of excision of megarectum in children with anorectal malformation. J Pediatr Surg. 2007;42(1):227–33.

14 Echo-Assisted Intersphincteric Autologous Microfragmented Adipose Tissue Injection to Control Fecal Incontinence in Children

Giovanni Parente, Valentina Pinto, Marco Pignatti, Neil Di Salvo, Simone D'Antonio, Michele Libri, and Mario Lima

14.1 Introduction

14.1.1 Fecal Incontinence

First of all it is important to highlight that it is possible to properly speak about continence only in children who completed toilet training: a report prepared for the Agency for Healthcare Research and Quality (AHRQ) of the U.S. Department of Health and Human Services (2006) suggested that most children between the ages of 18 and 30 months have the prerequisite skills to begin toilet training, and Brazelton et al. (1999) found that most children complete toilet training when 36 months old.

G. Parente (✉) · N. Di Salvo · S. D'Antonio · M. Libri
Pediatric Surgery Unit, IRCCS Sant'Orsola-Malpighi University-Hospital, Bologna, Italy
e-mail: niel.disalvo.mebi@pa.omceo.it; simone.dantonio@aosp.bo.it; michele.libri@aosp.bo.it

V. Pinto
Plastic Surgery Unit, IRCCS Azienda Ospedaliero-Universitaria di Bologna, Bologna, Italy
e-mail: valentina.pinto@aosp.bo.it

M. Pignatti
Plastic Surgery Unit, IRCCS Azienda Ospedaliero-Universitaria di Bologna, Bologna, Italy

DIMES, University of Bologna, Bologna, Italy
e-mail: marco.pignatti@unibo.it

M. Lima
Pediatric Surgery Unit, IRCCS Sant'Orsola, University-Hospital, University of Bologna, Bologna, Italy
e-mail: mario.lima@unibo.it

Fecal incontinence is defined as accidentally having bowel movements in patients who completed toilet training; it is a common condition in the pediatric population and can cause social unacceptance and embarrassment in both the parents and children the more the patients grow up.

Possible causes of fecal incontinence could be medically acquired conditions (like chronic constipations) or congenital anomalies (anorectal malformations, Hirschsprung disease, or spinal anomalies).

It is possible to distinguish the following:

True fecal incontinence: It regards children affected by anorectal malformations (ARM), Hirschsprung disease (HD), and spinal problems since these conditions prevent the patients from developing normal anatomical structures required for an efficient bowel control. Children with true fecal incontinence, in order to better evaluate the proper therapeutic protocol, should be further divided into patients with slow (**hypomotility**) or fast (**hypermotility**) bowel movements.

Pseudoincontinence or **encopresis**: It occurs in children with the ability to toilet train but who have developed a severe chronic constipation or in patients in whom the underlying disease predisposes them to develop constipation. In these patients, feces accumulate along the colorectal tract and the wateriest stools percolate through the hardest ones determining a fecal overflow that cannot be controlled even by a well-toilet-trained child without anatomical anomalies.

M. Lima, G. Ruggeri (eds.), *Ano-Rectal Endosonography and Manometry in Paediatrics*,
https://doi.org/10.1007/978-3-030-97668-2_14

Defecation is a result of a series of complex processes that requires three undamaged elements working together: **voluntary muscle control**, **sensation**, and colon's involuntary peristalsis (**motility**).

Voluntary muscles: The muscle complex of the perineum is under the conscious control of the baby: once stools arrive to the anal canal, the child is free to decide whether to relax the muscle complex (and then to defecate) or to hold the stool contracting the muscles.

The sphincter muscles can be weak in children born with ARM or spinal defects.

Sensation: The ability of the baby to feel the feces in the ampulla. The anal canal and ampulla, with their fine and reach innervation, are able to give the baby's brain extremely precise sensory information of their filling state to let him/her know when it is time to empty the ampulla.

Surgery and spinal defects could alter this perception making the brain no longer able to understand when the rectum is full of feces.

Motility: Continence issue could also arise from an altered speed of stool progression from the colon to the anal canal. In cases of hypomotility feces tend to gather in the ampulla determining a chronic constipation and then soiling to overflow. Hypermotility is a characteristic of postoperative patients who have the colon parts removed: these children experience a fast progression of feces, especially if watery, that leak out of the anus even in well-toilet-trained babies.

Treatment of fecal incontinence in children should always start from dietary changes and an accurate bowel management eventually powered by transanal irrigation systems.

In case of ineffectiveness of the previous mentioned methods, surgery could be taken into account.

Here the authors describe a novel echo-guided technique to enhance the continence ability of patients based on the intersphinceteric injection of adipose tissue derived from the patients themselves (autologous), also called anal lipofilling.

14.1.2 Adipose Tissue

Adipose tissue is a specialized connective tissue consisting of lipid-rich cells called adipocytes. As it comprises about 20–25% of total body weight in healthy individuals, the main function of adipose tissue is to store energy in the form of lipids (fat). Based on its location, fat tissue is divided into parietal (under the skin) and visceral (surrounding organs). Depending on the adipocyte morphology, there are two types of adipose tissue: white adipose tissue, mainly found in adults, and brown adipose tissue, mainly found in newborns.

Besides energy storing, fat tissue has several other important functions in the human body. These include thermal isolation, cushioning the organs, an endocrine role, and production of numerous bioactive factors.

Like every other tissue, adipose tissue consists of cells and extracellular matrix. The cells are the most abundant structural elements of this tissue, predominating over the small amount of extracellular matrix. The main cells that compose adipose tissue are the adipocytes. Besides adipocytes, several other cell types are present: fibroblasts, capillary endothelial cells, macrophages, and stem cells. These non-adipocyte cells collectively form the stromal vascular fraction, and their main function is to support and protect the adipose tissue. The extracellular matrix is produced by both adipocytes and stromal cells. It consists of a fine network of reticular fibers (type III collagen), whose function is to hold the cells in place. Adipose tissue is richly supplied with blood vessels and unmyelinated nerve fibers.

The abundance of the adipose tissue in the human body, its ease of obtaining, and its stem cell fraction make this tissue the best candidate for our technique (Fig. 14.1).

14.2 The Anal Lipofilling

Patients with fecal incontinence not responsive to medical treatments and bowel management, before more invasive surgical approach, may benefit from echo-assisted intersphincteric autologous microfragmented adipose tissue injection (anal lipofilling).

The ratio of this technique is that the injected fat tissue acts as a bulky agent increasing resting pressure as verified with anorectal manometry in various studies and thickens the anal sphincter as

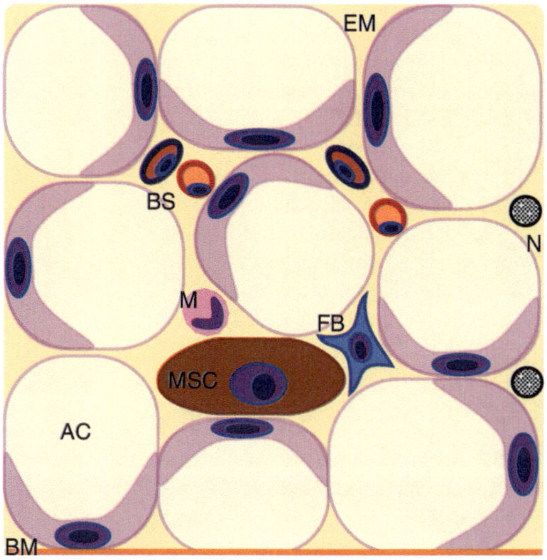

Fig. 14.1 Adipose tissue composition. *EM* Extracellular matrix, *BS* Blood supply, *N* Nerves, *M* Macrophages, *FB* Fibroblasts, *AC* Adipocyte, *MSC* Mesenchymal stem cells, *BM* Basal membrane

documented by anal endosonography by some authors.

Before its application in children, several studies reported its feasibility, safety, and encouraging results in adults.

The role of mesenchymal stem cells (MSCs) has still to be understood since in other disciplines (like orthopedics and cosmetics) a certain grade of tissue regeneration has been reported; therefore, MSCs may differentiate in order to implement the amount of sphincter muscle fibers but more studies are needed to clarify this point.

The main limits of autologous fat graft, also for anal lipofilling, remain to be the unpredictable viability and reabsorption of transferred fat. Therefore, the procedure needs to be repeated to maintain its efficacy.

14.2.1 Background of Fat Graft Application

Autologous adipose tissue transplantation is a well-established method with several clinical applications in plastic surgery.

It is based on the removal of fat graft from a donor site, its processing in order to make the tissue as suitable as possible for the recipient site, and the reimplantation of the processed tissue in a selected area.

The autologous adipose tissue has the main advantage of being completely biocompatible and easily available, also in pediatric patients, without morbidity for the donor site and without significant additional costs for the operating room. It is mainly used to correct a volumetric deficit, applicable in every part of the body. The transfer of autologous fat with filler purpose has been performed as whole grafts since the 1890s [1] and as injectable grafts since the 1920s [2] (breast reconstruction, facial and body contouring, post-traumatic treatments). Fat is the ideal filler: it is readily available and inexpensive to harvest, it is autologous and therefore lacks a host immune response, it is safe and noncarcinogenic, and it is acquired with a minimally invasive procedure. However, it is only within the last 15 years that the popularity of autologous fat graft for its regenerative purposes has increased within the plastic surgery and other surgical fields. Despite the well-known clinical advantages of the autologous fat graft, the main limit remains the unpredictable viability of transferred fat. During the years, several methods have been described to improve graft survival, better volume prediction, and regenerative capacity. Fat tissue is available in enough amount in most patients, also children, and can be harvested easily with a minimally invasive approach. Since the first description [3, 4], several studies have focused on the technical improvement and the optimization of the fat grafting technique described by Coleman. Simple lipoaspiration, gravity separation, Coleman fat centrifugation, washing, microfat, and nanofat techniques are the most cited and used approaches. Fat tissue offers a highly viable mesenchymal stem cell (MSC) population with optimal differentiation potential independent of the donor's age. The regenerative potential of adipose tissue-derived MSCs is similar to that reported in other tissues, but it is much more abundant compared with the widely used source of MSCs as bone marrow, and by their easy access [5]. A recent study compared the classical lipofilling technique with three commercial devices to obtain a fat

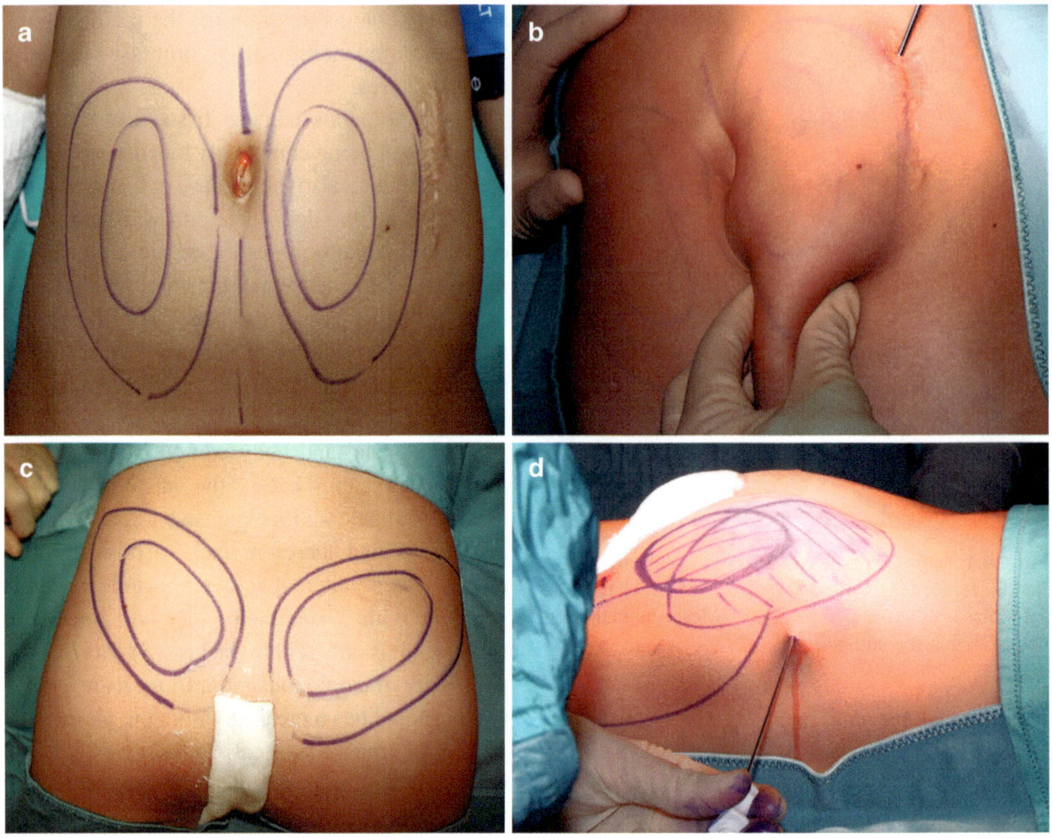

Fig. 14.2 Donor sites: (**a**) Abdominal region, (**b**) pinch test, (**c**) supragluteal area, (**d**) supragluteal fat graft harvesting

derivative enriched in MSCs, confirming that a greater amount of MSCs leads to better and more stable results [6]. Several studies showed greater tissue viability and a lower percentage of contaminants in fat tissue washed and filtrated within a closed system [7].

14.2.2 Surgical Procedure

The procedure of ano-lipofilling consists of three distinct phases performed in a single surgical time:

- Fat tissue harvesting
- Fat tissue processing
- Lipotreated tissue injection

14.2.3 Fat Tissue Harvesting

Each patient was examined preoperatively in orthostatic position with the pinch test, to precisely evaluate and draw the area to be lipoaspirated (Fig. 14.2).

The ideal donor site is represented by the abdomen, for different reasons.

Usually, abdominal area is chosen as the donor site because of its ease of access and availability; it is unique and median (for which we do not need to take bilateral sample as from the hips, thighs, and buttocks to avoid morpho-volumetric asymmetries). Further, the abdominal area as donor site allows to place the patient in supine position and easily converts it into a lithotomic position, needed for the injection time.

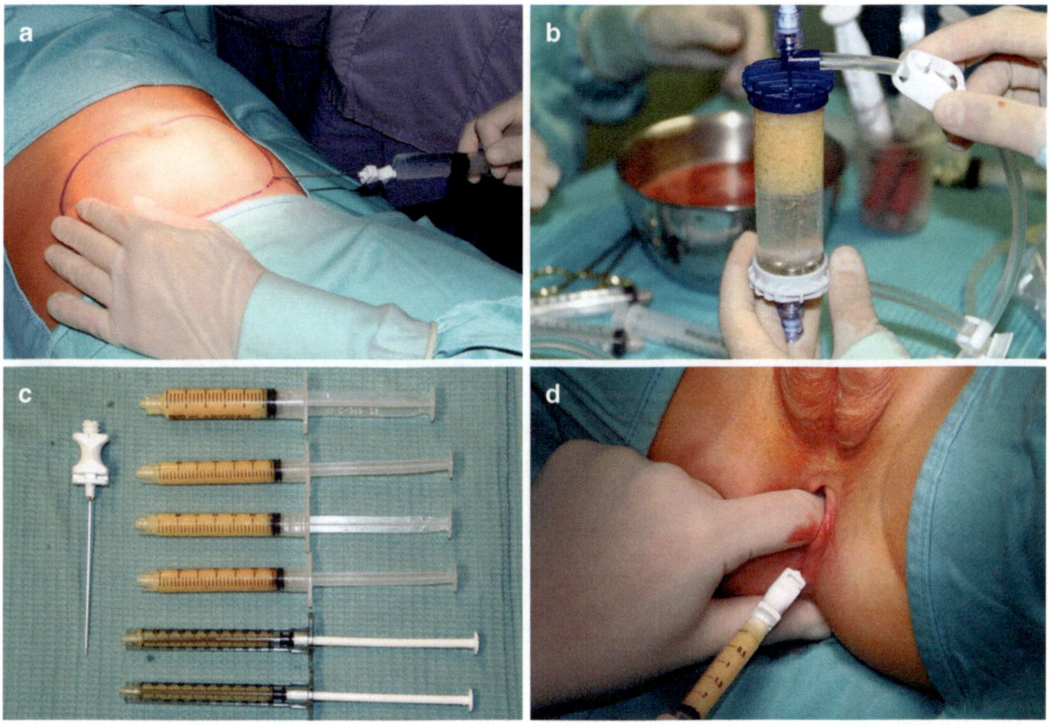

Fig. 14.3 Lipofilling procedure: (**a**) Local infiltration, (**b**) adipose tissue processing technique, (**c**) adipose tissue ready for injection, (**d**) intersphincteric adipose tissue injection

In very thin patients, the supragluteal region and inner knees are the best donors of adipose tissue but require bilateral sampling to preserve the contouring of the buttock; further they require to move the position with the patient's pronation-supination.

The first step is the infiltration in the donor site of a solution composed of epinephrine 2 mcg/mL in saline solution (1:500), using a 17-gauge cannula connected to the Luer Lock® syringe included in the kit (Fig. 14.3a).

After 10 min, lipoaspiration of a small amount of fat tissue is performed in a standardized fashion of the liposuction through a millimetric skin incision. Fat is harvested from the selected area using manual suction with a 20 mL VacLok® syringe and a 3 mm 13-gauge blunt cannula included in the Lipogems® kit (about 50/60 mL depending on the patient's body habitus). Sequential steps for fat tissue harvesting are reported in Table 14.1.

Table 14.1 Standard surgical liposuction: sequential steps

Standard surgical liposuction	
1	Skin incision with a 11 blade (peripheral area of the donor site) – *Hidden skin incision* – *Skin folds/umbilicus* – *Previous scars*
2	Infiltration of saline solution/epinephrine 2 mcg/mL in saline solution 1:500.000 in the selected area using disposable infiltration 17 G cannula (included in the Lipogems® kit)
3	Waiting for 10 min
4	Harvesting fat tissue in the selected area by using a blunt 13 G cannula with multiple elliptical holes (included in the Lipogems® kit) connected with a 20 cc self-locking Luer Lock syringe (VacLok®)
5	Transfer of the harvested fat tissue in the Lipogems® system

14.2.4 Fat Tissue Processing

Processing of the harvested fat is carried out with the Lipogems® device [5] (Fig. 14.3b), a closed,

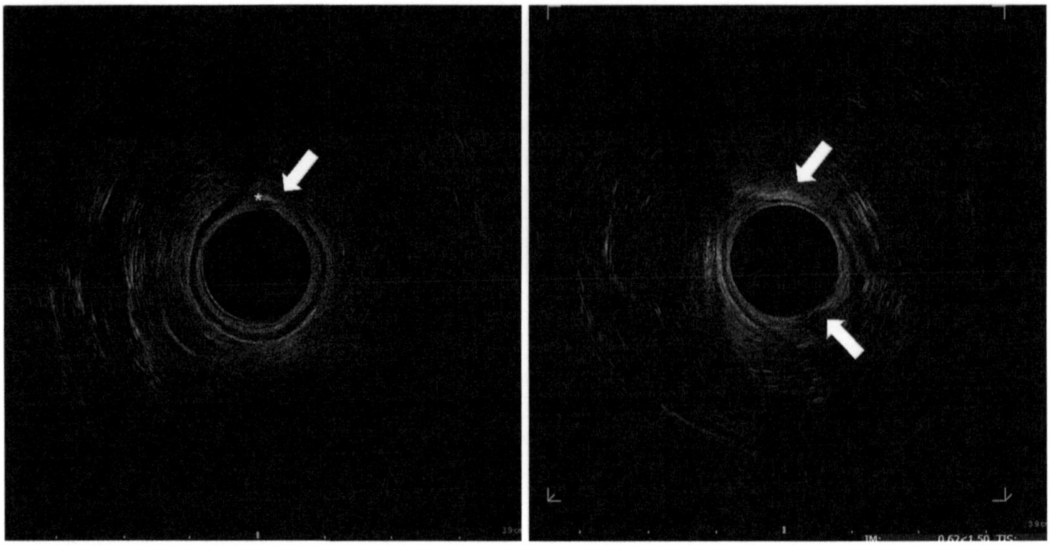

Fig. 14.4 Endosonographic axial sections of the middle anal canal. Asterisk: needle in the internal anal sphincter; arrow: injected fat tissue h 6 and 12

full-immersion, low-pressure cylindrical system, to obtain a gradual reduction in the adipose tissue clusters and remove impurities (blood, oil, and cell debris), producing an injectable fluid containing many pericytes/MSCs.

Lipogems® is a simple system patented in 2010 and clinically available since 2013, designed to harvest, process, and transfer refined adipose tissue, and is associated with great regenerative potential and optimal handling ability [5]. Thanks to this Italian system technology, fat tissue is microfragmented gently and washed from proinflammatory oil and blood residues, without enzymes or other additives. The processed fat is subjected to only slight mechanical forces, with no detrimental effects on the integrity of the stromal vascular components and on the tissue itself.

The processing procedure consists of five consecutive steps well described by the manufacturer; directions for use are supplied with the device. The system core is represented by a plastic cylinder containing five stainless steel marbles and connected to two hoses. The fat cluster reduction and purification are allowed by a mechanical system of metallic filters. The manual vertical shaking permits the steel spheres to emulsify and microfracture the adipose tissue. The fat is washed and emulsified, and adipose cluster size is gradually reduced to about 0.3–0.8 mm. This procedure will be repeated till washing saline solution will be clear and transparent. At this time, processed and purified fat graft, containing MSCs, is ready to be isolated and injected in the anal area.

In the resulting microfragmented fat, pericytes are retained within an intact stromal vascular niche and are ready to interact with the recipient tissue after injection, thereby becoming MSCs and starting the regenerative process (Fig. 14.3c).

14.2.5 Echo-Assisted Injection

Having completed the liposuction procedure and processed the adipose tissue, the patient is set in gynecological position and three injections of about 10 mL of adipose tissue are performed at three out of four quadrants of middle anal canal (Fig. 14.3d) under endoanal US guidance, precisely in the intersphincteric plane (Fig. 14.4).

As regards the echo-assistance, we use an anorectal 3D 2052 transducer (17 mm diameter, 13 MHz); endoanal ultrasound guarantees a more precise transplantation site.

Table 14.2 Donor-site treatment

Donor-site treatment
5.0 Prolene skin suture (single stitch): 1 week
Application of gauzes soaked with iced saline solution to reduce ecchymosis: up to postoperative dressing
Antibiotic therapy (amoxicillin): 5 days
Elastic and compressive dressing: 1 week
Heparan sulfate 1% cream to reduce ecchymosis

14.2.6 Postoperative Time

Immediate bulging and mechanical effect of the fat graft can be showed, while improvement of mucosal quality of the treated region is generally observed a few months after fat grafting. Compression dressing is applied on the donor site for 1 week. Antibiotic therapy (amoxicillin) is administered for 5 days. Heparan sulfate ointment is indicated to reduce ecchymosis (Table 14.2).

14.2.7 Complications

No hematomas, infections, and vascular or nervous injuries were reported in the treated children. No significant further surgical complications, from either the donor site or the injected site, were reported. Mild edema and bruising were frequent during the first postoperative week. Further possible described complications of this procedure could be visible irregularities of the liposuction site that are extremely rare considering the small amount of lipoaspiration and accurate preoperative evaluation. Major complications such as fat embolism syndrome present a low risk considering the small caliber of the blunt cannula used for lipoaspiration.

14.3 Authors' Experience

In our center we decided to apply this technique to patients with fecal incontinence due to ARM [8].

Table 14.3 International classification (Krickenbeck) for postoperative results in ARM

1. Voluntary bowel movements	
Feeling of urge, capacity to verbalize and hold the bowel movements	
1.0	No
1.1	Yes
2. Soiling/fecal incontinence	
2.0	No
2.1	Occasionally (once/twice a week)
2.2	Every day, no social problem
2.3	Constant, social problem
3. Constipation	
3.0	No
3.1	Manageable with diet
3.2	Requires laxatives
3.3	Resistant to diet and laxatives

We monitored the efficacy at follow-up using the international classification (Krickenbeck) for postoperative results in ARM (Table 14.3) [9].

We therefore treated four patients: they were all male, three rectourethral fistulas (75%), and one recto-perineal fistula (25%), and underwent the first anal lipofilling at a mean age of 13 ± 4.2 years (range: 8–17 years). All four patients suffered from total fecal incontinence; presented the same score at Krickenbeck scale, 1.0, 2.3, and 3.0; and needed an enema every day to help in bowel control.

The total number of procedures performed was 9 and the mean number of procedures per patient was 2.0 ± 1.3 (range: 1–4).

Three patients treated (75%) showed a significant improvement in continence with a KS score of 1.0, 2.1, and 3.0 after every single procedure and a reduction of the number of enemas from once daily to 1 or 2 weekly (Fig. 14.5).

One patient (25%) did not benefit from the anal lipofilling, and no changes on the KS were recorded before and after the procedure (Fig. 14.5).

The mean interval between the procedures in the ones who underwent more than one anal lipofilling was 343.8 days ±220.1 days (range 203–733 days).

No complications were recorded during and after the procedure.

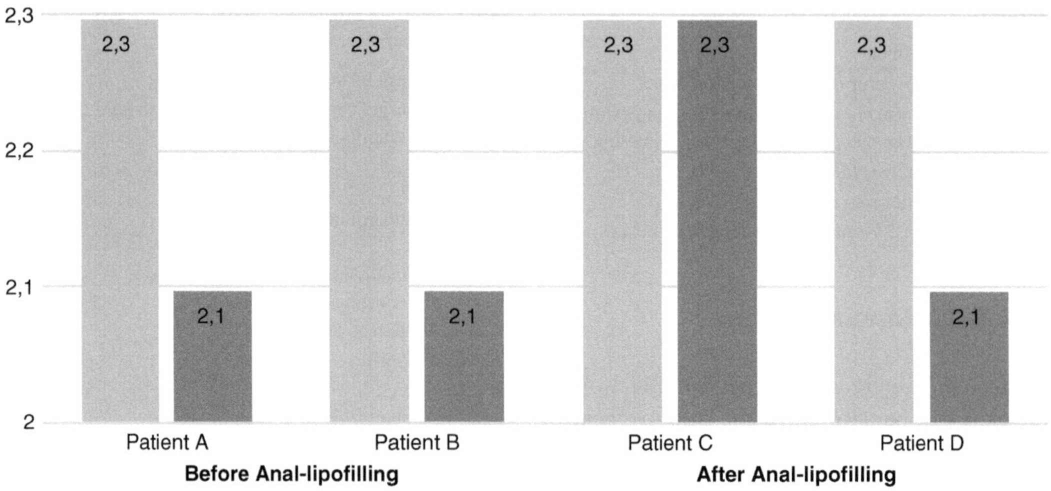

Fig. 14.5 Results of anal lipofilling on soiling reported as Krickenbeck scale (KS) score

14.4 Conclusions

The treatment of ARMs consists of a wide range of surgical and nonsurgical procedures [10].

Thanks to the optimal size of the clusters, allowing easy injection of the product and its regenerative capacity, fat graft and microfragmented fat have been tested and used safely in various clinical applications [11–16]. Fat grafting is a simple, effective, and reproducible technique, with a high satisfaction rate and few disadvantages or complications. According to our preliminary experience, fat tissue fits with authority into the portfolio of therapeutic options for patients with anorectal malformations. The microfragmentation technology (Lipogems®) improves and optimizes the natural properties of adipose tissue, without the use of enzymes, additives, or separation centrifugation. The opportunity to promote local tissue regeneration, thanks to mechanical MSC selection, makes local tissue more viscoelastic and represents a very promising approach for ARMs. The entire procedure is performed in a single surgical time and can be repeated without compromising the execution of further more invasive and traditional surgical procedures in the future.

References

1. Neuber F. Fettransplantation. Zentrabl Chir. 1893;22:66.
2. Miller CG. Cannula implants and review of implantation techniques in esthetic surgery. Chicago: Oak Press; 1926.
3. Coleman SR. Structural fat grafting: more than a permanent filler. Plast Reconstr Surg. 2006;118(3S):108S–20.
4. Coleman SR. Structural fat grafting. Aesth Surg J. 1998;18(5):386–8.
5. Tremolada C, Colombo V, Ventura C. Adipose tissue and mesenchymal stem cells: state of the art and Lipogems® technology development. Curr Stem Cell Rep. 2016;2:304–12. https://doi.org/10.1007/s40778-016-0053-5.
6. Domenis R, Lazzaro L, Calabrese S, Mangoni D, Gallelli A, Bourkoula E, et al. Adipose tissue derived stem cells: in vitro and in vivo analysis of a standard and three commercially available cell-assisted lipotransfer techniques. Stem Cell Res Ther. 2015;6:2.
7. Zhu M, Cohen SR, Hicok KC, Shanahan RK, Strem BM, Yu JC, et al. Comparison of three different fat graft preparation methods: gravity separation, centrifugation, and simultaneous washing with filtration in a closed system. Plast Reconstr Surg. 2013;131(4):873–80.
8. Parente G, Pinto V, Di Salvo N, D'Antonio S, Libri M, Gargano T, Catania VD, Ruggeri G, Lima M. Preliminary study on the Echo-assisted Intersphincteric autologous microfragmented adipose tissue injection to control fecal incontinence in children operated for anorectal malformations. Children

(Basel). 2020;7(10):181. https://doi.org/10.3390/children7100181.
9. Holschneider A, et al. Preliminary report on the international conference for the development of standards for the treatment of anorectal malformations. J Pediatr Surg. 2005;40(10):1521–6.
10. Kyrklund K, et al. Social morbidity in relation to bowel functional outcomes and quality of life in anorectal malformations and Hirschsprung's disease. Eur J Pediatr Surg. 2018;28(6):522–8.
11. Barfod KW, et al. Treatment of osteoarthritis with autologous and microfragmented adipose tissue. Dan Med J. 2019;66(10):A5565.
12. Cestaro G, De Rosa M, Massa S, Amato B, Gentile M. Intersphincteric anal lipofilling with microfragmented fat tissue for the treatment of faecal incontinence: preliminary results of three patients. Wideochir Inne Tech Maloinwazyjne. 2015;10(2):337–41.
13. Jeong H, Hwang SH, Kim HR, et al. Effectiveness of autologous fat graft in treating fecal incontinence. Ann Coloproctol. 2019;35(3):144–51. https://doi.org/10.3393/ac.2018.09.15.
14. Boric I, et al. A 24-month follow-up of the effect of intra-articular injection of autologous microfragmented fat tissue on proteoglycan synthesis in patients with knee osteoarthritis. Genes. 2019;10(12)
15. Mautner K, et al. Functional outcomes following microfragmented adipose tissue versus bone marrow aspirate concentrate injections for symptomatic knee osteoarthritis. Stem Cells Transl Med. 2019;8(11):1149–56.
16. Laureti S, et al. Refractory complex Crohn's perianal fistulas: a role for autologous microfragmented adipose tissue injection. Inflamm Bowel Dis. 2020;26(2):321–30.

15 Endoanal Ultrasonography in Male and Female Urogenital Malformations

Mario Lima, Michela Maffi, Neil Di Salvo, Giovanni Ruggeri, and Marzia Vastano

Abbreviations

ARM	Anorectal malformations
3D EUS	3-Dimensional endorectal ultrasonography
DSD	Disorders of sex development
EUS	Endorectal ultrasonography
MRI	Magnetic resonance imaging
PU	Prostatic utricle
S/A	Stromal/area
TAS	Transabdominal ultrasounds
TVS	Transvaginal ultrasounds
VCUG	Voiding cystourethrogram

15.1 Introduction

Genital malformations in children are usually studied by abdominal ultrasound and MRI. From a historical point of view, abdominal ultrasound and MRI have been considered the gold standard approach for the diagnosis, but in the last few years, 3D endoanal US has been gaining an increasing interest in the pediatric world.

M. Lima · M. Maffi · N. Di Salvo
G. Ruggeri (✉) · M. Vastano
Pediatric Surgery Unit, IRCCS Sant'Orsola University-Hospital, University of Bologna, Bologna, Italy
e-mail: mario.lima@unibo.it; michela.maffi@aosp.bo.it; niel.disalvo.mebi@pa.omceo.it

Endoanal ultrasonography in children is currently used to detect malformations or injuries of anal channel, but it constitutes also an alternative way to explore the other pelvic structures. Transrectal way is already well known in adults, both in males and females.: in males, mainly to investigate the prostate, and in females for internal genitalia when the vaginal US is not applicable. In this chapter we explore possible applications of 3D endoanal sonography in the detection of urogenital malformations and other conditions in children.

15.2 Endoanal Ultrasonography in Females

Little girls and adolescents may present different gynecological problems. In women, vaginal examination and transvaginal US (TVS) are routinely used to investigate internal genitalia. In pediatric age, the presence of intact hymen makes this modality intolerable and culturally not acceptable. Transabdominal ultrasounds (TAS) are useful to explore internal genitalia, but when the visualization is suboptimal, endorectal US can provide a further way of investigation [1, 2].

EUS can be performed also with transvaginal probes, but some discomfort during the movement of the probe is reported [1].

M. Lima, G. Ruggeri (eds.), *Ano-Rectal Endosonography and Manometry in Paediatrics*, https://doi.org/10.1007/978-3-030-97668-2_15

The use of 2D and 3D endoanal ultrasounds (3D EUS) performed with appropriate probes may remove the possible discomfort.

15.2.1 Ovaries

3D EUS has been used to investigate ovaries in different conditions.

In polycystic ovarian syndrome, EUS allows for objective determination of ovarian volume, number and size of follicles, stromal volume, overall blood flow, ovarian total and stromal area, and S/A ratio.

Sun and Fu [3] reported 3D EUS combined with TAS to be superior to TVS in the diagnosis and evaluation of polycystic ovarian syndrome in adolescents.

EUS is also able to detect adnexal masses and to characterize them according to the most used scores [2, 4].

15.2.2 Uterovaginal Malformations

Congenital defects of vaginal canalization are rare and include vaginal septa and vaginal agenesis alone or associated with uterine agenesis (Rokitansky syndrome). Transrectal ultrasonography provides an accurate map of the pelvic organs showing the urethra and bladder anteriorly, rectum posteriorly, retrohymenal fovea caudally, and pelvic peritoneum cranially. In this malformation there are a lot of anatomical variants and it is important that the surgeon obtains as much anatomical information as possible preoperatively. To plan an adequate correction of these malformations, it is important to know if the patient has a complete vaginal agenesis or partial atresia, the extension of atresic tract, the presence of a uterus or a uterine remnant and a cervix, and the presence and entity of associated hematocolpos. Transrectal ultrasonography produced more than other exams a picture that corresponded perfectly with the real anatomical situation and could become a further diagnostic tool in the assessment of vaginal canalization defects as it provides images as good as MRI [1, 5, 6].

In postpubertal girls with a functioning uterus, vaginal obstructive malformations are diagnosed during the investigations for amenorrhea or after the onset of cyclic pelvic pain. The main finding in these cases is hematocolpos. EUS can identify hematocolpos as a fluid vaginal collection and estimate its dimensions and volume. Different conditions may cause vaginal obstruction. Imperforate hymen is easier to diagnose as an inspection of external genitalia is sufficient to visualize bulging bluish mass. EUS in this case can be useful to confirm diagnosis and exclude other associated malformations [6–8].

Vaginal septum can also be detected by EUS that can also estimate its extension and provide a differential diagnosis with partial or complete vaginal agenesis.

Atresic vagina can be visualized as a complete absence of structure between rectum and urethra or as a hypoechoic, non-cavitated band.

The detection of functioning or rudimental uterus and presence of endometrium and cervix are mandatory in a diagnostic workup. Also in these cases, EUS can investigate uterine features guiding the treatment options, in particular if preserving or removing uterine rudiment [5].

15.2.3 Particular Cases

15.2.3.1 OHVIRA Syndrome

OHVIRA syndrome, also known as Herlyn-Werner-Wunderlich syndrome, is a rare condition characterized by uterus didelphis, obstructed hemivagina, and ipsilateral renal agenesis. It belongs to Müllerian anomalies with associated mesonephric malformation. These patients are usually asymptomatic in prepubertal age, but in literature, some cases of early onset of this syndrome under 5 years are reported [9]. After the onset of menses, they begin to complain of cyclic pelvic pain and development of abdominal mass. The diagnosis can be delayed due to the normal appearance of external genitalia and the presence of menses from the patent uterus and vagina [10].

The transrectal US shows abnormal distention of the obstructed hemivagina and can estimate the distance between the obstructed vagina and

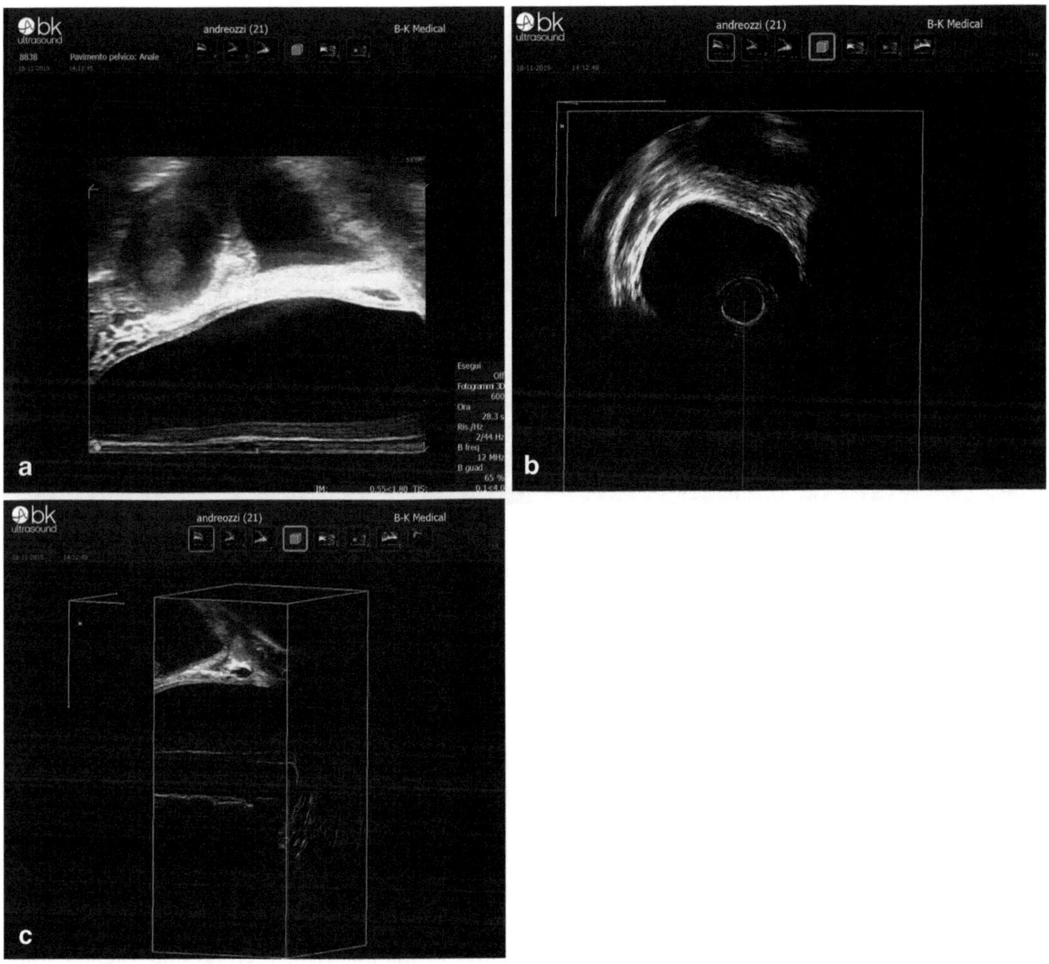

Fig. 15.1 Case of an 11-year-old female patient affected by OHVIRA syndrome. She has cyclic abdominal pain. US revealed renal agenesis and obstructed hemivagina. Diagnostic workup was completed with 3D EUS and MRI. Both the 2D (**a**) and 3D EUS (**b**, **c**) performed before the debulking demonstrate the extension of the obstructed hemi-system that surrounds the rectum for almost 180°

the perineal plane. This information given by endosonography is almost the same as that of MRI. 2D scans allow to study the internal genital anatomy, and then it is possible to reconstruct a 3D image of the anal canal and vagina, uterus, adnexa, and bladder (Fig. 15.1).

15.2.3.2 Genital Anomalies and Anorectal Malformations

The association between anorectal malformations (ARM) and gynecological anomalies is well known and ranges from 26 to 39%. Different malformations can be present involving vagina, uterus, or both. The incidence of genital malformation in ARM grows with the severity of the anorectal malformation being more frequent in the cloaca than in the less complex forms. Moreover, patients with genital malformations present more frequently associated urological anomalies [11–16]. The presence should be investigated before or during definitive treatment of ARM. Vaginoscopy can detect vaginal septa and if an abdominal approach is necessary, internal genitalia can be explored. Nevertheless, in some cases, gynecological anomalies can be misdiagnosed at birth. During the follow-up of ARMs, EUS can be useful both to check the anal

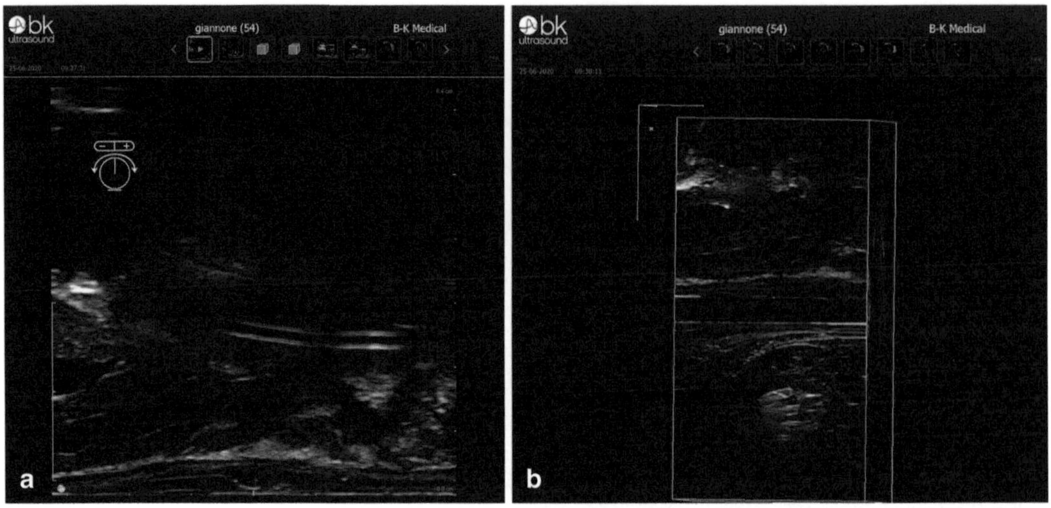

Fig. 15.2 Case of an 11-year-old female affected by persistence of cloaca corrected during the first years of age. She came to our attention for vaginal stenosis and utero-vaginal septum. Both the 2D (**a**) and 3D EUS (**b**) revealed stenotic vaginal channel and dilated vagina with septum

channel and the internal genitalia to discover any undiagnosed malformations and to check the results of corrective interventions already performed (Fig. 15.2).

15.2.4 Further Applications: Urethral Injuries

Normal sonographic findings of the urethra are continuity of the wall, patent lumen, smoothness of inner surface, and uniformity of muscular layer echo.

Urethral injuries are usually partial and longitudinal urethral wall laceration and are often associated to vaginal laceration due to the proximity of the two structures.

In these cases, EUS can be performed with the patient in lateral position with a comfortably full bladder. Both the static and the voiding phases can be observed for a better definition of trauma entity.

Urethral hematoma appears as a mixed echo mass surrounding the urethra and usually has a spontaneous regression at follow-up.

Urethral strictures are characterized by a dilatation of the proximal urethra in the voiding phase. Damaged urethra also appears with harsh cavity surface.

Scars are detected as irregular hyper- or hypoechoic zone localized in the retropubic space and surrounding urethra and vagina.

Urethrovaginal fistula appears as a discontinuity of the posterior urethral and vaginal walls with a narrow hypoechoic or anechoic zone between them.

In *urethral rupture*, urethral wall is discontinuous, urethral muscular layer shows irregular echoes in the injured tract, sometimes the distal urethra is not visible, and some scars can appear [17].

15.3 Endoanal Ultrasonography in Males

The transrectal US is a well-known approach to investigate the prostate and other structures localized in posterior urethra [18]. In adults, prostatic pathologies are quite frequent, so the benefit of this approach is high. In children, pathologies involving posterior urethra are few; nevertheless, EUS allows a good evaluation of pelvic structures also in males.

15.3.1 Prostate

In children, prostatic masses are rare and most are prostatic rhabdomyosarcoma [19, 20], but in literature carcinoid tumors and prostatic benign hyperplasia are also reported [19, 21].

EUS can be used to identify prostatic masses that cause an enlargement of the normal gland. Moreover, ultrasound guidance can be used to perform biopsies. Currently the experience in this field reported in the literature is limited only to case reports, but we can think that this approach may have a future.

15.3.2 Prostatic Utricle

Prostatic utricle (PU) is an enlarged diverticulum localized in the posterior urethra and derived by the persistence of Müllerian structures or decreased androgenic stimulation of the urogenital sinus. It is a rare condition with an estimated prevalence of 5% in urologic patients and an incidence of 1% in autopsy [22, 23]. PU is mostly associated to hypospadias while only rarely is an isolated condition. Most of the PU is asymptomatic but it can also be manifested by urinary infections, urinary retention, epididymitis, stone formation, and post-void dribbling. These symptoms are usually related to compressive effect of the PU on surrounding structures [24].

Ultrasounds are the first-line imaging method for PU providing diagnosis in most of the cases. Also VCUG can reveal a PU as the utricular chamber is filled with contrast medium. MRI usually completes the diagnostic workup. Nevertheless, transrectal ultrasounds have been applied in the diagnosis and treatment of PU in adults and in some pediatric cases [22, 25]. In Fig. 15.3, one case is reported in which 3D EUS has been applied in the diagnostic workup.

15.3.3 Hypothesis for Further Applications in Children

Unfortunately, the literature on the use of 3D EUS in pediatric age for the study of urogenital pathologies is still enormously limited. However, considering the quality of the images that can be obtained, it is possible to hypothesize their use in some other conditions.

A conceivable use in males could be the follow-up of the posterior urethral valves where it would be possible to study the posterior urethra without the use of ionizing radiation.

We have applied 3D EUS in a patient affected by exstrophy-epispadias complex, partially corrected at birth. It was possible to highlight the absence of puborectalis muscle, posterior urethra, and bladder neck (Fig. 15.4).

Another possible use could concern patients with disorders of sex development (DSD), to evaluate the presence of a urogenital sinus or Müllerian remnants. Also in this case, given the use of ultrasound in females with malformations of the internal genitalia, it is conceivable that comparable results would be obtained in terms of quality and usefulness of the images.

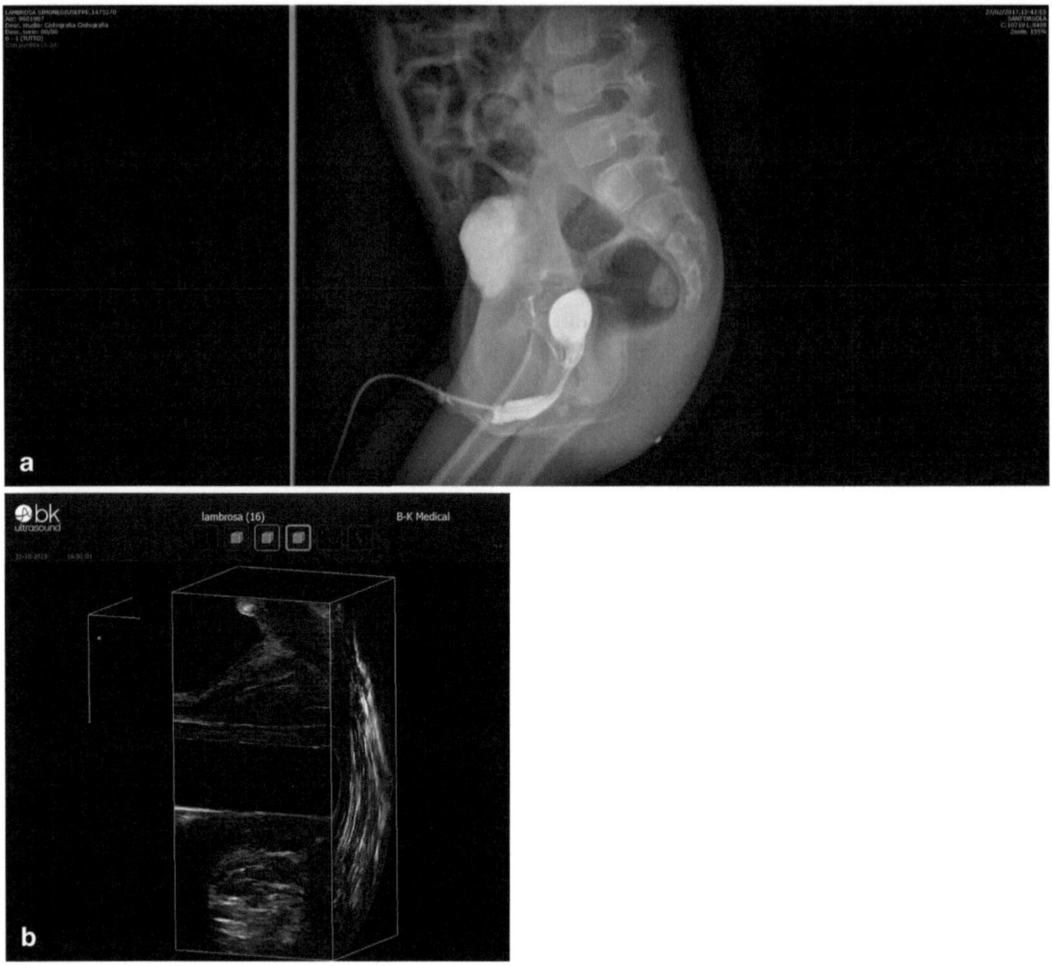

Fig. 15.3 Case of a 7-year-old boy affected by Russell-Silver syndrome, with proximal hypospadias. 3D EUS allowed to point out, as the VCUG, (**a**) the presence of enlarged prostatic utricle between the rectal anterior wall and the urinary way (**b**) (blue arrow)

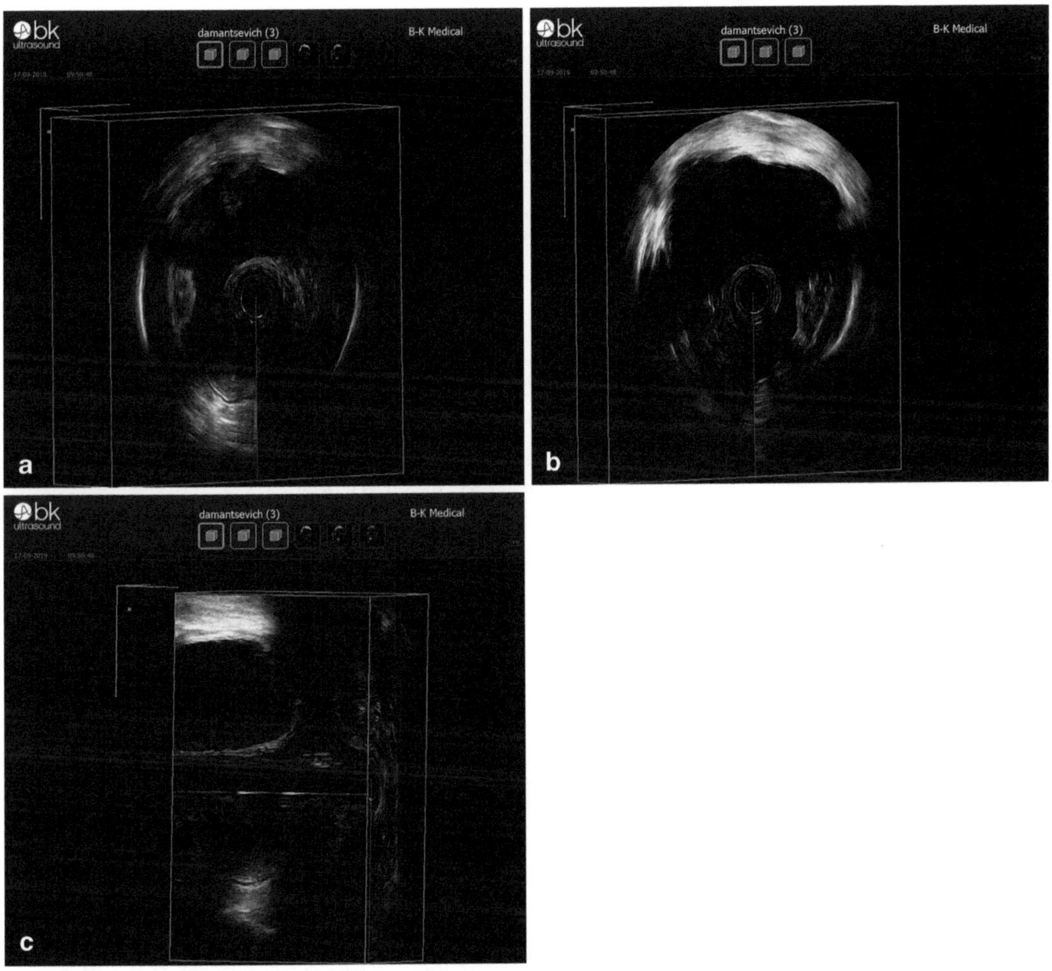

Fig. 15.4 Case of a 12-year-old boy affected by exstrophy-epispadias complex, partially corrected at birth (bladder closure). Both the axial (**a**, **b**) and longitudinal (**c**) scans show the absence of puborectalis muscle, posterior urethra, and bladder neck

References

1. Güdücü N, Sidar G, İşçi H, Yiğiter AB, Dünder İ. The utility of transrectal ultrasound in adolescents when transabdominal or transvaginal ultrasound is not feasible. J Pediatr Adolesc Gynecol. 2013;26(5):265–8. https://doi.org/10.1016/j.jpag.2013.04.004. Epub 2013 Jul 10.
2. Timor-Tritsch IE, Monteagudo A, Rebarber A, Goldstein SR, Tsymbal T. Transrectal scanning: an alternative when transvaginal scanning is not feasible. Ultrasound Obstet Gynecol. 2003;21(5):473–9. https://doi.org/10.1002/uog.110.
3. Sun L, Fu Q. Three-dimensional transrectal ultrasonography in adolescent patients with polycystic ovarian syndrome. Int J Gynaecol Obstet. 2007;98(1):34–8. https://doi.org/10.1016/j.ijgo.2007.02.024. Epub 2007 Apr 24.
4. Meys EMJ, Jeelof LS, Achten NMJ, Slangen BFM, Lambrechts S, Kruitwagen RFPM, Van Gorp T. Estimating risk of malignancy in adnexal masses: external validation of the ADNEX model and comparison with other frequently used ultrasound methods. Ultrasound Obstet Gynecol. 2017;49(6):784–92. https://doi.org/10.1002/uog.17225.
5. Fedele L, Portuese A, Bianchi S, Zanconato G, Raffaelli R. Transrectal ultrasonography in the assessment of congenital vaginal canalization defects. Hum Reprod. 1999;14(2):359–62. https://doi.org/10.1093/humrep/14.2.359.

6. Anguenot JL, Ibecheole V, Salvat J, Campana A. Hematocolpos secondary to imperforate hymen, contribution of transrectal echography. Acta Obstet Gynecol Scand. 2000;79(7):614–5.
7. Kushnir O, Garde K, Blankstein J. Rectal sonography for diagnosing hematocolpometra. A case report. J Reprod Med. 1997;42(8):519–20.
8. López-Rasines G, Abascal F, Calabia A, Rico M, Cavero A. Transrectal sonography in the assessment of vaginal pathology: a preliminary study. J Clin Ultrasound. 1998;26(7):353–6. https://doi.org/10.1002/(sici)1097-0096(199809)26:7<353::aid-jcu5>3.0.co;2-8.
9. Angotti R, Molinaro F, Bulotta AL, Bindi E, Cerchia E, Sica M, Messina M. Herlyn-Werner-Wunderlich syndrome: An "early" onset case report and review of Literature. Int J Surg Case Rep. 2015;11:59–63. https://doi.org/10.1016/j.ijscr.2015.04.027. Epub 2015 Apr 24.
10. Tuna T, Estevão-Costa J, Ramalho C, Fragoso AC. Herlyn-Werner-Wunderlich syndrome: report of a prenatally recognised case and review of the literature. Urology. 2019;125:205–9. https://doi.org/10.1016/j.urology.2018.12.022. Epub 2018 Dec 28.
11. Fanjul M, Lancharro A, Molina E, Cerdá J. Gynecological anomalies in patients with anorectal malformations. Pediatr Surg Int. 2019;35(9):967–70. https://doi.org/10.1007/s00383-019-04513-1. Epub 2019 Jul 3.
12. Levitt M, Bischoff A, Breech L, Peña A. Rectovestibular fistula-rarely recognized associated gynecologic anomalies. J Pediatr Surg. 2009;44(6):1262–7. https://doi.org/10.1016/j.jpedsurg.2009.02.046.
13. Peña A, Bischoff A. Rectovestibular fistula. In: Peña A, Bischoff A, editors. Surgical treatment of colorectal problems. 1st ed. New York: Springer; 2015. p. 205–24.
14. Breech L. Gynecologic concerns in patients with anorectal malformations. Semin Pediatr Surg. 2010;19(2):139–45. https://doi.org/10.1053/j.sempedsurg.2009.11.019.
15. Breech L. Gynecologic concerns in patients with cloacal anomaly. Semin Pediatr Surg. 2016;25(2):90–5. https://doi.org/10.1053/j.sempedsurg.2015.11.006.
16. Rintala R. Congenital cloaca: long-term follow-up results with emphasis on outcomes beyond childhood. Semin Pediatr Surg. 2016;25(2):112–6. https://doi.org/10.1053/j.sempedsurg.2015.11.011.
17. Ying T, Li Q, Shao C, Zhu Z, Feng L, Hu B. Value of transrectal ultrasonography in female traumatic urethral injuries. Urology. 2010;76(2):319–22. https://doi.org/10.1016/j.urology.2009.11.044. Epub 2010 Feb 13.
18. Watanabe H, Igari D, Tanahashi Y, Harada K, Saitoh M. An evaluation of the function of new special equipment for transrectal ultrasonotomography. Tohoku J Exp Med. 1976;118(4):387–92. https://doi.org/10.1620/tjem.118.387.
19. Yağmur I, Turna B, Tekin A, Akıncıoğlu E, Sarsık B, Ulman İ. Benign prostatic hyperplasia: case report of a 17-year-old. J Pediatr Urol. 2016;12(4):267.e1–4. https://doi.org/10.1016/j.jpurol.2016.04.031. Epub 2016 May 28.
20. Kuru TH, Roethke MC, Nyarangi-Dix J, Okouoyo S, Stockklausner C, Schenk JP, Debus J, Roth W, Teber D, Pahernik S, Schlemmer HP, Hohenfellner M, Hadaschik BA. Pediatric case report on magnetic resonance imaging/transrectal ultrasound-fusion biopsy of rhabdomyosarcoma of the bladder/prostate: a new tool to reduce therapy-associated morbidity? Urology. 2013;81(2):417–20. https://doi.org/10.1016/j.urology.2012.10.043.
21. Llorens de Knecht E, Guadarrama Vega S, Donate G, Palou Redorta J, Bujons Tur A. Robot-assisted radical prostatectomy due to a primary carcinoid prostatic tumor in a three-year-old child: a case report. Urology. 2019;133:216–8. https://doi.org/10.1016/j.urology.2019.08.006. Epub 2019 Aug 12.
22. Liu B, He D, Zhang D, et al. Prostatic utricles without external genital anomalies in children: our experience, literature review, and pooling analysis. BMC Urol. 2019;19:21. https://doi.org/10.1186/s12894-019-0450-z.
23. Wang W, Wang Y, Zhu D, Yan P, Dong B, Zhou H. The prostatic utricle cyst with huge calculus and hypospadias: a case report and a review of the literature. Can Urol Assoc J. 2015;9(5–6):E345–8. https://doi.org/10.5489/cuaj.2381.
24. Hester AG, Kogan SJ. The prostatic utricle: an under-recognized condition resulting in significant morbidity in boys with both hypospadias and normal external genitalia. J Pediatr Urol. 2017;13(5):492.e1–5. https://doi.org/10.1016/j.jpurol.2017.01.019. Epub 2017 Mar 1.
25. Coppens L, Bonnet P, Andrianne R, de Leval J. Adult müllerian duct or utricle cyst: clinical significance and therapeutic management of 65 cases. J Urol. 2002;167(4):1740–4.

Endoanal Ultrasonography: Use in Adults

16

Dajana Cuicchi and Gilberto Poggioli

The endoanal ultrasound (EAUS) technique was first described by Bartram and Law in 1989 using a BrÜel & Kjaer ultrasound scanner type 1846 with a 7 MHz rotating endoanal probe having a focal length of 3 cm which generated images in the axial plane only [1]. Subsequently, improvements in the technology of ultrasound scanners and transducers (multifrequency probes and 3-dimensional endosonography) allowed a more complete and precise evaluation of the anal canal and distal rectum in the coronal, axial and sagittal planes. Thus, EAUS has rapidly become a popular technique for imaging the low rectum and anal canal in patients with a variety of anorectal diseases. It is useful in the staging, restaging and follow-up of rectal and anal cancers; it also plays a major role in evaluating benign diseases, such as faecal incontinence and perianal fistulas or abscesses. It can sometimes be useful in cases of obstructed defaecation. It is most applicable in the setting of a doctor's office where immediate treatment decisions can be made based on the information provided [2].

The aim of this chapter is to describe the role of ultrasound in studying the anal canal in order to assess functional and neoplastic pathologies in adults. For the discussion of suppurative processes of the anorectal region, there is a dedicated chapter to which we refer. Attention will therefore focus on three disorders: faecal incontinence, obstructed defaecation and anal cancer.

16.1 Faecal Incontinence

Faecal incontinence (FI) is a symptom common to various disorders which can alter one or more of the mechanisms involved in maintaining continence to the extent of invalidating the compensation provided by the remaining unimpaired structures [3]. The cause of incontinence is often multifactorial [4]. Careful clinical assessment of the patient and anorectal instrumental diagnostics provide useful information in defining the contribution of the individual etiopathogenetic factors, determining the severity of the problem and establishing the impact of FI on the quality of life [5].

Endoanal ultrasound (EAUS) makes it possible to observe the complexity of the mechanisms involved in faecal incontinence, showing sphincter defects in patients considered to be affected by neuropathic incontinence, or not revealing significant anatomical alterations in patients strongly suspected of suffering from sphincter damage or showing muscular lesions in completely asymptomatic subjects (occult anal sphincter defect).

D. Cuicchi (✉) · G. Poggioli
Surgery of the Alimentary Tract, Department of Digestive Medicine & Surgery, IRCCS Azienda Ospedaliero Universitaria di Bologna, Alma Mater Studiorum University of Bologna, Bologna, Italy
e-mail: dajana.cuicchi@aosp.bo.it; gilberto.poggioli@unibo.it

M. Lima, G. Ruggeri (eds.), *Ano-Rectal Endosonography and Manometry in Paediatrics*,
https://doi.org/10.1007/978-3-030-97668-2_16

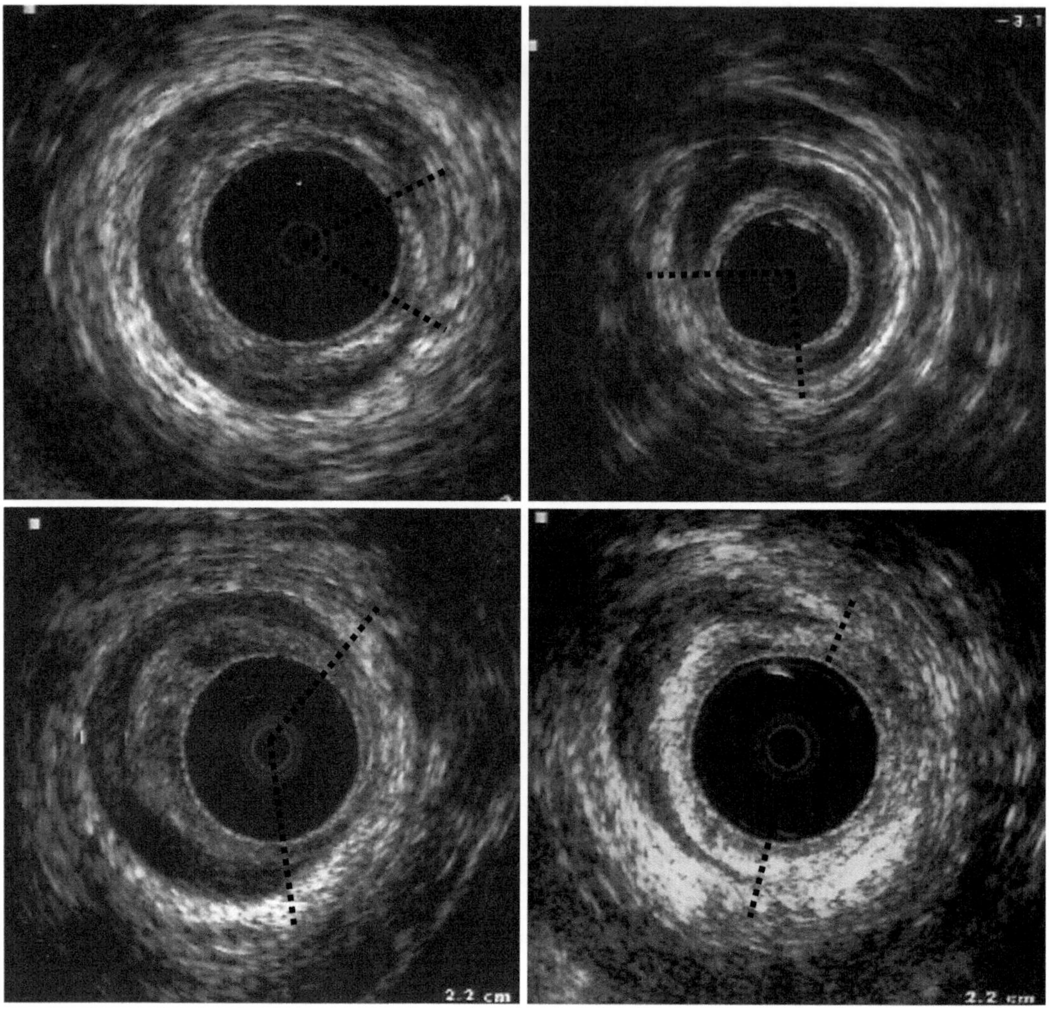

Fig. 16.1 Internal anal sphincter defects

Endoanal ultrasound allows evaluation of the thickness and anatomical integrity of the internal anal sphincter (IAS) and the external anal sphincter (EAS), and detection of the presence of scar tissue with a high degree of accuracy [6]. This method has been validated by both in vivo and in vitro studies, and has demonstrated an almost 100% sensitivity and specificity in determining the topography of the sphincter defects [7, 8]. The importance of this method in diagnosing faecal incontinence is due to the fact that, in the majority of cases of incontinence, sphincter defects are responsible; EAUS is currently the gold standard for sphincter evaluation [9].

At EAUS, tears of the IAS are seen as an interruption of the hypoechoic ring, sometimes associated with compensatory hypertrophy of the remaining muscle portion; the scar is an area of amorphous texture which is usually hyperechoic (Fig. 16.1). They are mostly iatrogenic following internal lateral sphincterotomy, fistulotomy, hemorrhoidectomy, anal divulsions and transanal interventions; however, they can also be caused by penetrating trauma or complete rectal prolapse. In EAS defects, the scar areas appear as amorphous areas which are hypoechoic or mixed echogenic areas (Fig. 16.2). The most frequent cause of EAS defects is obstetric

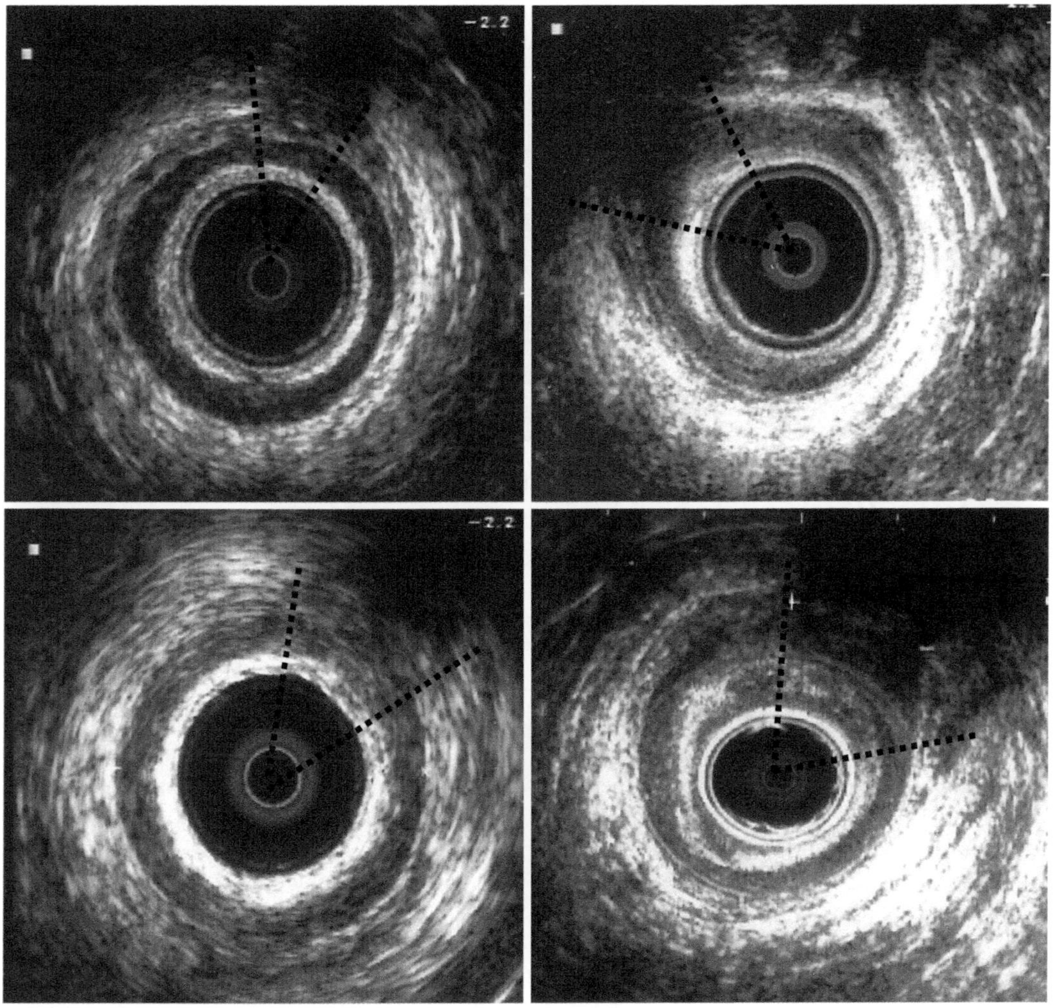

Fig. 16.2 External anal sphincter defects

trauma which is the most frequent cause of incontinence in women (Fig. 16.3); the defect is located anteriorly. The fragmentation of the IAS and/or the EAS is defined as two or more fragments in the axial plane. It occurs in patients after multiple surgeries, grade III–IV obstetric anal sphincter injury and perineal trauma. During EAUS, the number, circumferential extent (radial angle in degrees or in hours of the clock) and longitudinal extent (proximal, distal or full length) of the defect should be reported. Two EAUS-based scoring systems have been proposed to define the severity of anal sphincter damage, both of them in women with obstetric anal sphincter injuries. In Starck's scoring classification, the absence of an anal sphincter defect is indicated by 0 and a defect >180° involving the entire length and depth of both sphincters corresponds to 16 [10]. Norderval et al. reported a simplified system, including fewer categories without recording partial defects of the IAS [11]. The maximal score of 7 denoted defects in both the EAS and the IAS exceeding 90° in the axial plane and involving more than half of the length of each sphincter. Both scoring systems have demonstrated good correlation between the extent of the sphincter defects and the degree of FI.

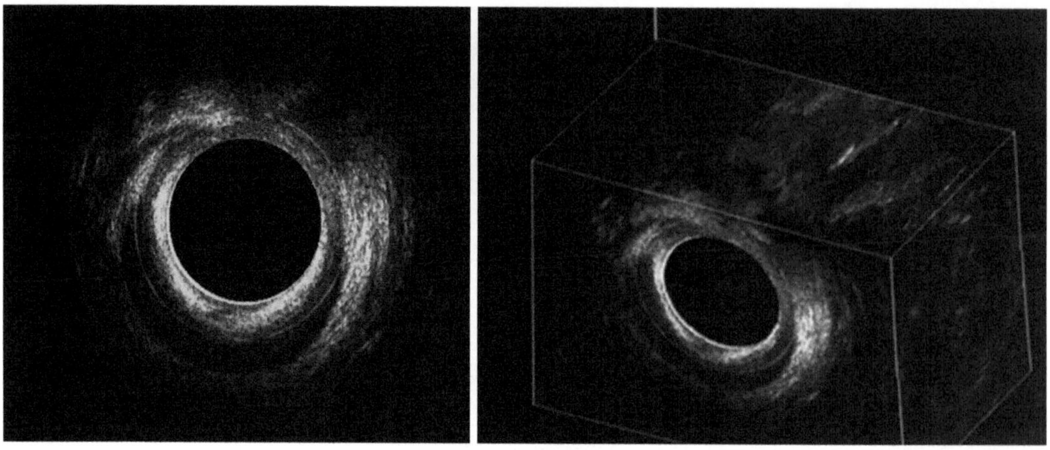

Fig. 16.3 Type II obstetrical anal sphincter injuries at clinical evaluation and type IIIc at endoanal ultrasound

Perineal body measurement improves visualisation of the anterior sphincter lesions in females. A perineal body thickness of 10 mm or less is considered abnormal whereas 10–12 mm is associated with a sphincter defect in one-third of patients, and those with 12 mm or more are unlikely to harbour a defect unless they had previously undergone reconstructive perineal surgery [12].

Three-dimensional (3D) EAUS allows length, thickness, area and volume measurement. It also improves the percentage of identifying sphincter defects. Christensen et al. evaluated the differences between 3D and 2D EAUS in visualising damage to the anal sphincters [13]. Three-dimensional EAUS improved diagnostic confidence as compared to 2D; in fact, agreement between the two observers who evaluated the images was better when the EAUS was performed in 3D (98.2% vs. 87.9%, respectively).

One possible pitfall during EAUS is that it can simulate sphincter tears resulting in a 5–25% of false positives in identifying sphincter defects. The most common normal anatomical finding which can be interpreted as a sphincter defect is the physiological absence of the EAS in women in the anterior quadrant at the level of the high anal canal and the puboanal muscles which constitute the most medial part of the puborectal muscle and the anococcygeal ligament present in the posterior quadrant at the level of the middle anal canal, and which connects the EAS and the puborectal muscle to the coccyx. Finally, at the level of the passage between the middle and low anal canal, an asymmetric termination can be found in the IAS which can simulate a defect.

In the absence of sphincter defects EAUS is able to identify an additional subgroup of patients affected by passive faecal incontinence caused by a uniform and pathological thinning of the IAS (<2 mm in diameter in patients over the age of 50). This rare condition can be secondary to conditions of systemic sclerosis (scleroderma) or primary disease (Fig. 16.4). The primary degeneration of the smooth muscle cells of the IAS was described for the first time in 1997 [14].

A considerable limitation of this method is the difficulty in identifying the neurogenic atrophy of the EAS. Since neurological damage causes the loss of functional tissue and its consequent replacement with adipose tissue, it is, in fact, difficult to identify the interface separating the EAS from the fat of the ischioanal space and, consequently, the precise measurement of the thickness of the striated muscle [6]. Williams et al. carried out a prospective study on 25 patients without US-detectable lesions of the EAS, 16 of whom were incontinent, having reduced sphincter pressure, and 9 of whom were healthy controls, performing magnetic resonance imaging (MRI) with an endoanal coil and electromyography [15]. The authors showed that the loss of definition of the US interface at the outer margin of the striated muscle and the thinning of the IAS identified

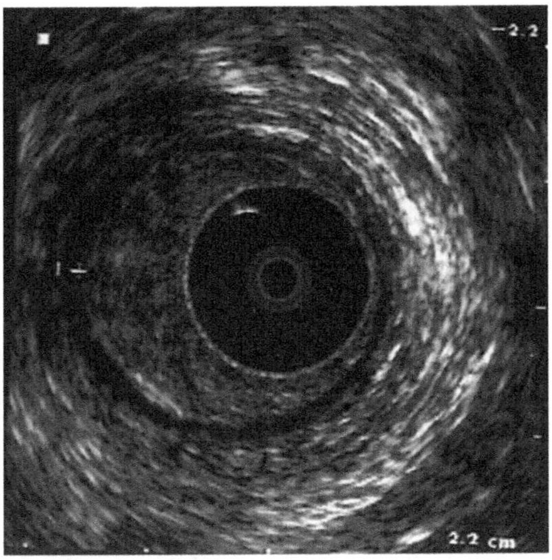

Fig. 16.4 Primary degeneration of the internal anal sphincter

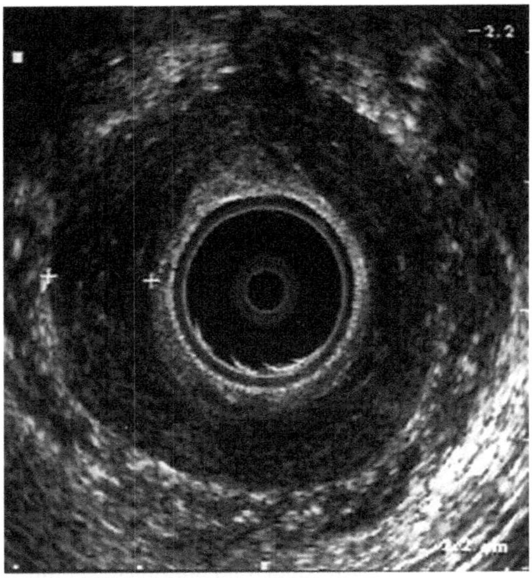

Fig. 16.5 Hereditary myopathy at endoanal ultrasound

EAS atrophy with a positive predictive value of 74%. It is important to recognise atrophy because it is associated with a poor clinical outcome of sphincter repair. If EAS atrophy is suspected, endoanal MRI should be performed. At MRI, the loss of muscular fibres and their replacement with adipose tissue are accurately defined, thanks to the different intensities of the signals emitted by the two tissues. Considering histological examination as the reference gold standard, endoanal MRI has a sensitivity and specificity of 89% and 94%, respectively, in determining the presence or absence of atrophy of the external sphincter [16].

Endoanal ultrasound is also useful for evaluating the results of the surgical treatment of patients with faecal incontinence. In cases of symptom persistence after sphincteroplasty, EAUS may initially reveal undiagnosed EAS atrophy or identify the persistence of the EAS defect due to technically imperfect repair, suture failure or other defects initially misunderstood. After bulking agent injection, it can also show a dislocation of the implants [17].

The high degree of definition of the anal canal and its accuracy in characterising any sphincter defects make EAUS an indispensable examination in the assessment of patients with incontinence; it can also confirm the integrity of the sphincters in cases of suspected neuropathic incontinence. The relationship between structural damage and clinical symptoms must nevertheless be clarified by carrying out additional tests in view of the numerous factors which can be involved in causing faecal incontinence. The presence of a sphincter defect does not necessarily mean that it is the only cause of the faecal incontinence.

16.2 Obstructed Defaecation

The only disorder associated with obstructed defaecation in which EAUS plays a leading diagnostic role is hereditary internal anal sphincter myopathy, a rare condition clinically characterised by constipation and proctalgia fugax, and at ultrasound by a pathological thickening of the internal anal sphincter of up to 1 cm (Fig. 16.5) [18]. Other conditions in which EAUS can be used, but with less diagnostic value, are solitary rectal ulcer syndrome, intussusception and rectal prolapse. All three disorders are associated with a thickening of the internal anal sphincter, albeit to a lesser extent with respect to hereditary myopathy [19, 20]. In solitary rectal ulcers, ultrasound can detect other features, such as hyperechoic fibrotic areas and multiple cysts in the submucosa, in the absence of both an interruption of the layers of the rectal wall and regional lymph node involvement [21]. Although, in histologically doubtful cases, these findings could suggest the benign nature of the lesion, biopsy must be repeated for the definitive exclusion of a mucinous carcinoma. Finally, in patients with rectal prolapse EAUS can reveal damage of the internal anal sphincter secondary to prolapse or other causes [20].

16.3 Anal Cancer

The high accuracy in defining the structures of the anal canal together with the characteristic appearance of gastrointestinal neoplasms at ultrasound study has made EAUS fundamental in the different phases of the management of anal cancer: (1) it plays a leading role in staging and therefore in treatment planning (surgery vs. radiation therapy-chemotherapy [RT-CHT]); (2) in cases where it is necessary to complete the therapy with brachytherapy, it can optimise the positioning of the needles; (3) at the end of the treatment, it can provide information regarding the treatment response and contribute to the early identification of cases of persistence of the disease; and (4) finally, in the follow-up, it can detect a loco-regional relapse at an early stage.

At EAUS, anal cancer appears as a non-homogeneous hypoechoic area with irregular edges well defined from the surrounding structures. Endoanal ultrasound provides accurate information regarding the circumferential and longitudinal extension of the tumour and the depth of infiltration into adjacent tissues. Different ultrasound staging systems have been proposed [22–24]. Boman et al. proposed a system in which T1 lesions are limited to the submucosa, and invasion into the "sphincter complex" is considered a T2 lesion [22]. Bartram et al. differentiated tumours that invaded the IAS from those that penetrated into the EAS but they did not make any allowance for tumours limited to the submucosa [23]. The most used staging system is that proposed by Tarantino and Bernstein in 2002 in which five main categories are considered: tumours confined to the submucosa (uT1); tumours which infiltrate the IAS (uT2a) or EAS (uT2b); neoplasms which extend through the sphincter complex into the adipose tissue of the ischioanal fossa (uT3); and tumours which infiltrate adjacent organs or anatomical structures (uT4) (Fig. 16.6) [24]. This classification helps to differentiate early lesions (uT1) treatable by local surgery from more advanced lesions which require chemoradiotherapy (CRT). Therefore, this staging system appears to have a greater clinical impact on treatment decisions than the TNM system proposed by the American Joint Committee on Cancer in which the categories of the T parameter are established in relation to the size of the neoplastic lesion (T1 ≤2 cm; T2 2–5 cm; T3 >5 cm; T4, infiltration of contiguous organs) [25]. Some studies have shown that primary tumour size does not represent an independent prognostic factor. In a multicentric study, it was shown that, in tumours classified as T1-2N0 according to ultrasound staging, the complete response rate to multimodal treatment was significantly higher than that recorded in the same categories (T1-2N0) identified according to the TNM system (94% vs. 80%, respectively, $p = 0.008$) [26]. Furthermore, in the multivariate analysis, the ultrasound T stage was found to be the only predictor of survival. The lymphatic pathway of tumour spread occurs through the perirectal, internal iliac and inguinal lymph nodes. Using endorectal ultrasound, it is only possible to evaluate the possible involvement of the perirectal lymph nodes. The size criterion is not reliable since approximately 2/3 of the positive lymph nodes are smaller than 5 mm. The morphological criteria are more reliable; a round lesion, with regular margins and with echogenicity similar to that of the primary tumour, is highly suggestive of the involvement of a secondary lymph node. The accuracy of EAUS and MRI was compared by Otto et al. who found good concordance between the two diagnostic techniques, suggesting that the former was superior for the detection of superficial tumours while the latter was needed for N staging since EAUS cannot visualise the inguinal or iliac lymph nodes which are outside the field of vision [27].

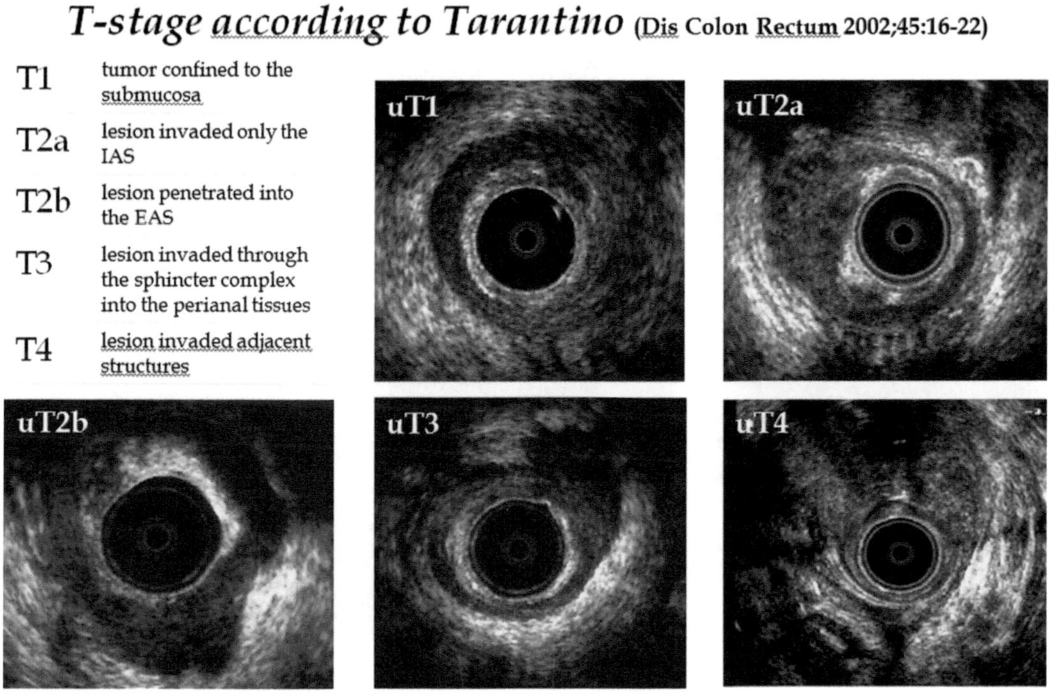

Fig. 16.6 T-stage according to Tarantino classification

If interstitial brachytherapy is indicated after the completion of CRT, EAUS can define the topography of the neoplastic residue, in this way optimising the positioning of radioactive sources (needles).

Endoanal ultrasound is also useful in the period following multimodal treatment, even though tumour response after CRT is often difficult to determine due to oedema and scar tissue at the location of the tumour as a result of the radiation treatment. However, some general rules may be useful; in the case of tumour response, the tumour is initially seen at EAUS as an area of mixed echogenicity which tends to disappear over time, that is, to be less clearly identifiable than the surrounding structures. In the case of a notable fibrotic reaction, the tumour evolves in the hyperechoic area; clearly hypoechoic areas should be considered as potential foci of the persistence of the tumour disease. Therefore, to correctly define the response to treatment, it is important to follow the morphological evolution of the ultrasound findings over time (Fig. 16.7). The optimal timing for assessing treatment response has not yet been determined. However, 6–8 weeks after the end of the treatment seems to be too short a period in the majority of cases due to the persistence of the effects of the therapy. Some authors have found that 16–20 weeks after radiation treatment is a sufficient time frame to allow for resolution of the oedema and accurate ultrasound imaging [24]. Ultrasound repeated over time allows careful selection of the cases in which it is necessary to perform biopsies owing to the suspicion of persistence of the disease.

Finally, EAUS can be used in the follow-up of patients treated with multimodal therapy to identify any neoplastic relapses at an early stage. However, its place in this setting may be debatable. While evolutionary criteria and technological innovations (colour Doppler, 3D EAUS) facilitate the early identification of relapses, there are doubts regarding the real added value of ultrasound as compared to clinical evaluation alone [28–30].

In conclusion the ultrasound staging of anal cancer is reliable from the prognostic point of view and is capable of influencing multimodal

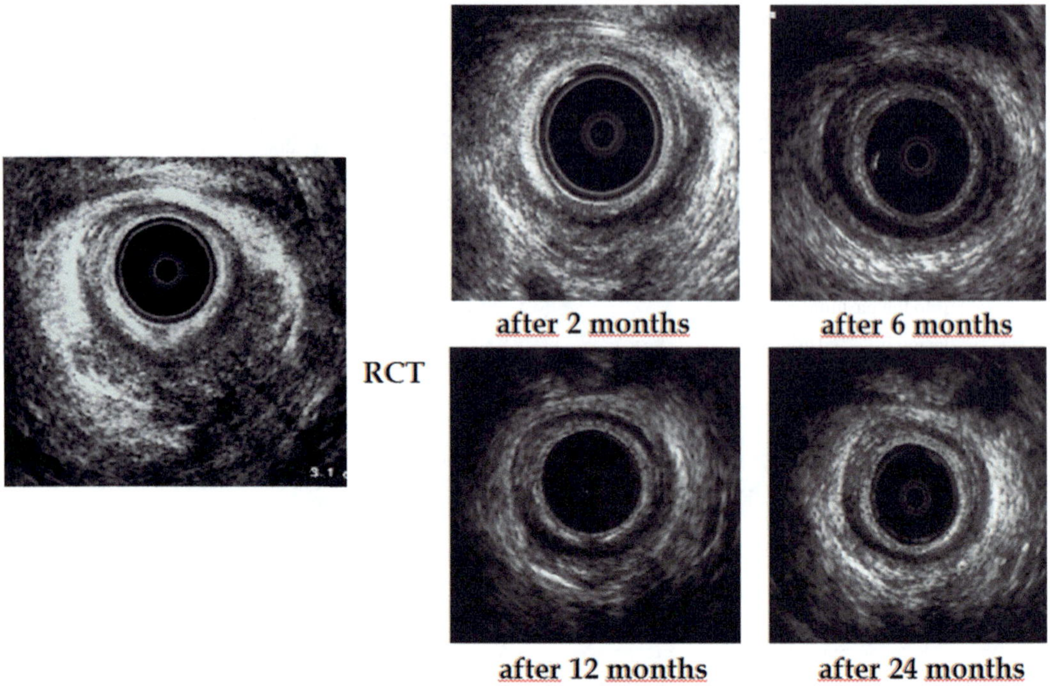

Fig. 16.7 Morphological evolution of the ultrasound findings of anal cancer over time after radiochemotherapy (RCT)

treatment in all its phases; the repetition of EAUS at the end of multimodal therapy allows identifying non-responsive patients as candidates for biopsy examination; eliminating doubt is the role of EAS in the follow-up.

References

1. Law PJ, Bartram CI. Anal endosonography: technique and normal anatomy. Gastrointest Radiol. 1989;14:349–53.
2. Schaffzin DM, Wong WD. Surgeon-performed ultrasound: endorectal. Ultrasound Surg Clin N Am. 2004;84:1127–49.
3. Rao SSC. Diagnosis and management of fecal incontinence. Am J Gastroenterol. 2004;2:1585–604.
4. Sun WM, Donnely TC, Read NW. Utility of a combined test of anorectal manometry, electromyography and sensation in determining the mechanism of idiopathic fecal incontinence. Gut. 1992;33:807–13.
5. Bharucha AE. Fecal incontinence. Gastroenterology. 2003;124:1672–85.
6. Bartram CI. Functional anorectal imaging. Abdom Imaging. 2005;30:195–203.
7. Sultan AH, Nicholls RJ, Kamm MA, et al. Anal endosonography and correlation with in vivo and in vitro anatomy. Br J Surg. 1993;80:508–11.
8. Law PJ, Kamm MA, Bartram CI. A comparison between electromyography and anal endosonography in mapping external anal sphincter defects. Dis Colon Rectum. 1990;33:370–3.
9. Tjandra JJ, Dykes SL, Kumar RR, et al. Practice parameters for the treatment of fecal incontinence. Dis Colon Rectum. 2007;50:1497–507.
10. Starck M, Bohe M, Valentin L. Results of endosonographic imaging of the anal sphincter 2-7 days after primary repair of third- or fourth-degree obstetric sphincter tears. Ultrasound Obstet Gynecol. 2003;22:609–15.
11. Norderval S, Markskog A, Røssaak K, et al. Correlation between anal sphincter defects and anal incontinence following obstetric sphincter tears: assessment using scoring systems for sonographic classification of defects. Ultrasound Obstet Gynecol. 2008;31:78–84.
12. Zetterström JP, Mellgren A, Madoff RD, et al. Perineal body measurement improves evaluation of anterior sphincter lesions during endoanal ultrasonography. Dis Colon Rectum. 1998;41:705–13.
13. Christensen AF, Nyhuus B, Nielsen MB, et al. Three dimensional anal endosonography may improve

diagnostic confidence of detecting damage to the anal sphincter complex. Br J Radiol. 2005;78: 308–11.
14. Vaizey CJ, Kamm MA, Bartram CI. Primary degeneration of the IAS as a cause of passive faecal incontinence. Lancet. 1997;349:612–5.
15. Williams AB, Bartram CI, Modhawadia D, et al. Endocoil magnetic resonance imaging quantification of external anal sphincter atrophy. Br J Surg. 2001;88:853–9.
16. Briel JW, Zimmerman DD, Stoker J, et al. Relationship between sphincter morphology on endoanal MRI and histopathological aspects of the external anal sphincter. Int J Color Dis. 2000;15:87–90.
17. de la Portilla F, Reyes-Diaz ML, Maestre MV, et al. Ultrasonographic evidence of Gatekeeper™ prosthesis migration in patients treated for faecal incontinence: a case series. In J colorectal dis. 2017;32:437–40.
18. Kamm MA, Hoyle CH, Burleigh DE, et al. Hereditary internal anal sphincter myopathy causing proctalgia fugax and constipation. A newly identified condition. Gastroenterology. 1991;100:805–10.
19. Marshall M, Halligan S, Fortheringham T, et al. Predictive value of internal anal sphincter thickness for diagnosis of rectal intussusception in patients with solitary rectal ulcer syndrome. Br J Surg. 2002;89:1281–5.
20. Bartram CI, DeLancey JOL. Imaging pelvic floor disorders. Berlin Heidelberg New York: Springer; 2003.
21. Cola B, Cuicchi D, Dalla Via B, et al. Endosonographic pattern of solitary polypoid rectal ulcer. Tech Coloproctol. 2005;9:71–2.
22. Boman BM, Moertel CG, O'Connell MJ, et al. Carcinoma of the anal canal: a clinical and pathologic study of 188 cases. Cancer. 1984;54:114–25.
23. Bartram CI, Burnett SJ. Atlas of anal endosonography. Oxford: Butterworth-Heinemann; 1991.
24. Tarantino D, Bernstein MA. Endoanal ultrasound in the staging and management of squamous-cell carcinoma of the anal canal: potential implications of a new ultrasound staging system. Dis Colon Rectum. 2002;45:16–22.
25. NCCN clinical practice guidelines in oncology: anal carcinoma. National Comprehensive Cancer Network available at https://www.nccn.org/professionals/physician_gls/pdf/anal.pdf. Version 1.2020-November 19, 2019.
26. Giovannini M, Bardou VJ, Barclay R, et al. Anal carcinoma: prognostic value of endorectal ultrasound (ERUS). Results of a prospective multicenter study. Endoscopy. 2001;33:231–6.
27. Otto SD, Lee L, Buhr HJ, et al. Staging anal cancer: prospective comparison of transanal endoscopic ultrasound and magnetic resonance imaging. J Gastrointest Surg. 2009;13:1292–8.
28. Magdeburg B, Fried M, Meyenberger C. Endoscopic ultrasonography in the diagnosis, staging and follow up of anal carcinomas. Endoscopy. 1999;31:359–64.
29. Lund JA, Sundstrom SH, Haaverstad R, et al. Endoanal ultrasound is of little value in follow up of anal carcinomas. Dis Colon Rectum. 2004;47:839–42.
30. Drudi FM, Giovagnorio F, Raffetto N, et al. Transrectal ultrasound color Doppler in the evaluation of recurrence of anal canal cancer. Eur J Radiol. 2003;47:142–8.

GPSR Compliance

The European Union's (EU) General Product Safety Regulation (GPSR) is a set of rules that requires consumer products to be safe and our obligations to ensure this.

If you have any concerns about our products, you can contact us on ProductSafety@springernature.com

In case Publisher is established outside the EU, the EU authorized representative is:

Springer Nature Customer Service Center GmbH
Europaplatz 3
69115 Heidelberg, Germany

Batch number: 10731385

Printed by Printforce, the Netherlands